SYSTEMATICS OF THE ONOCLEOID FERNS

BY

ROBERT M. LLOYD

UNIVERSITY OF CALIFORNIA PRESS
BERKELEY · LOS ANGELES · LONDON
1971

University of California Publications in Botany

Volume 61

Approved for publication May 28, 1971
Issued November 24, 1971

University of California Press
Berkeley and Los Angeles
California

University of California Press, Ltd.
London, England

ISBN: 0-520-09411-5
Library of Congress Catalog Card No.: 72-170330

printed in the united states of america

CONTENTS

SYSTEMATICS OF THE ONOCLEOID FERNS

BY

ROBERT M. LLOYD*

INTRODUCTION

THE CLARIFICATION OF relationships among fern genera and their species has been accomplished in the past primarily by character correlation of one or more morphological features considered by the authors to be important in determining relationship. The most definitive morphological work to date has been done on the more primitive fern families and there is now general agreement as to their evolutionary relationships within the Pteridophyta. Until recently classification systems of ferns were dominated by the ideas of Diels (1902), who created the very large family Polypodiaceae for the most advanced ferns having a gradate sorus and a vertical, interrupted annulus. Recent trends have been toward splitting the Polypodiaceae *sensu* Diels into several families (Alston, 1956; Ching, 1940; Copeland, 1947; Holttum, 1947; Wagner, 1969). A comparison of these systems indicates considerable disagreement over the constitution of families and their relationships to one another. The basis for these discrepancies has been, to a large extent, the lack of comparative morphological information.

Within the ferns there are many problematic genera and groups of genera whose relationships have not yet been clarified. This study concerns one of the more perplexing and persistent enigmas in the Filicales: the origin and relationship of the onocleoid ferns. This group comprises three genera: ***Onoclea,*** a monotypic genus of the eastern United States and Asia (pl. 1); ***Onocleopsis,*** a recently described genus from southern Mexico and Guatemala (pls. 3 and 4); and ***Matteuccia,*** a complex of three species found in Europe, North America, and Asia (pl. 5). The Asian species of this last genus were segregated into the genus *Pentarhizidium* by Hayata (1928).

In this study I hope to throw some light on two of the basic problems within the onocleoid complex: (1) their relations to one another; and (2) their relations within the dennstaedtioid-aspidioid complex of Filicales. As a guideline for generic comparisons I have chosen to follow the system presented by Wagner (1969), since it is basically similar to those of Copeland (1947) and Holttum (1947) and takes into account much of the more recently obtained data on ferns. Wagner's scheme divides the Filicales into four major groups: (1) the evolutionarily isolated Osmundaceae; (2) the gymnogrammoid ferns; (3) the polypodioid ferns; and (4) the dennstaedtioid-aspidioid ferns (fig. 1).

The gymnogrammoid ferns are a complex of five families grouped together on the basis of such morphological features as presence of a gymnogrammoid sorus (sporangia spread over the top of the veins or arranged linearly along the margin of the pinna), lack of a true indusium, and possession of tetrahedral spores without a well-developed perispore. The major families in the group are Adiantaceae, Vittariaceae, and Schizaeaceae *sensu lato*. The Adiantaceae contains genera which form a large part of Copeland's Pteridaceae.

* Present Address: Department of Botany, Ohio University, Athens, Ohio 45701.

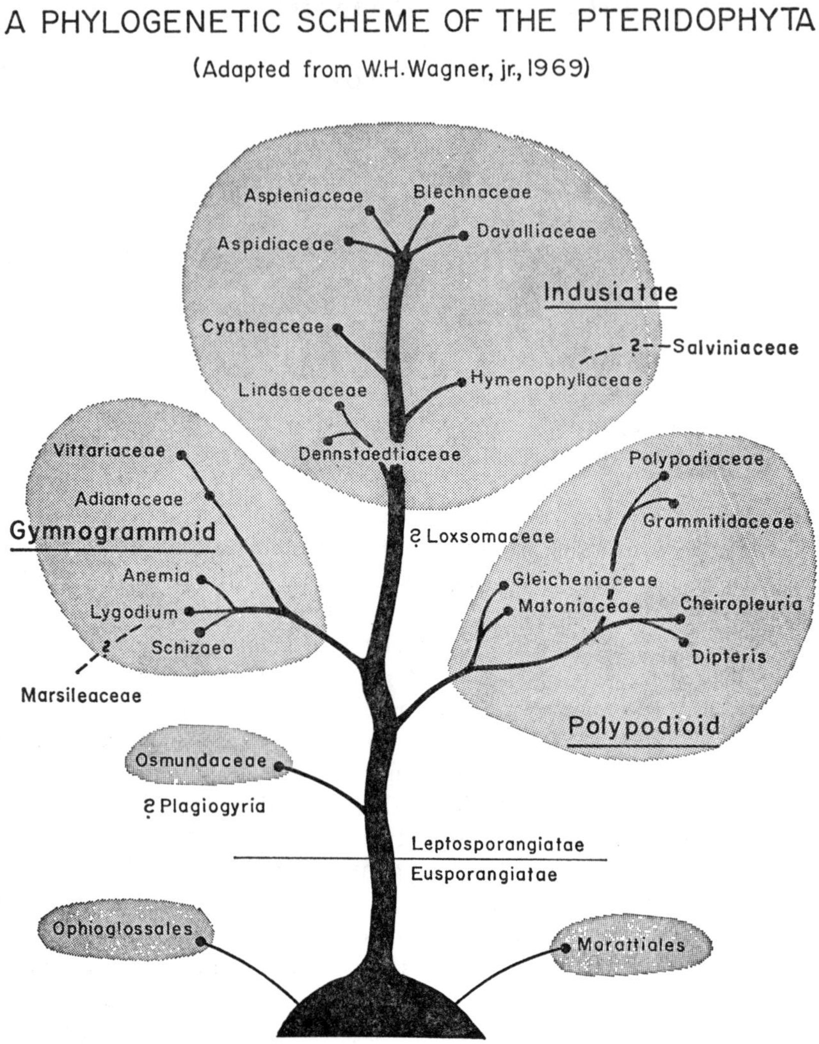

Fig. 1. A phylogenetic scheme of the higher Pteriodophyta, adapted from Wagner, 1969.

The polypodioid ferns comprise several families including Gleicheniaceae, Grammitidaceae, and Polypodiaceae, all of which have a pectinate frond structure bearing punctate to mammiliform exindusiate sori and tetrahedral or bilateral spores which lack a well-developed perispore.

The dennstaedtioid-aspidioid line, the largest and most advanced group of ferns, is characterized by the presence of a euindusium covering the sorus and bilateral (or rarely trilete) spores commonly bearing a perispore. The basic type of indusium is the cup-shaped structure found in the Dennstaedtiaceae. From this have evolved the various asymmetrical indusia found in the aspidioid family complex, which includes Aspidiaceae, Blechnaceae, Aspleniaceae, and Davalliaceae. This aspidioid complex contains about 2,000 species in 94 genera.

I have used evidence from comparative morphology of all stages of the life-cycle of the onocleoid ferns, as well as chromatographic evidence. Information from studies on the reproductive biology of the three genera (Klekowski and Lloyd, 1968) as well as experiments on spore germination (Lloyd and Klekowski, 1970) has also been used.

Acknowledgments

I am especially grateful to Dr. Robert Ornduff for his guidance throughout the course of this study and to Drs. Lincoln Constance and Elizabeth Cutter for their critical comments on the manuscript. I would also like to thank Dr. W. H. Wagner, Jr., of the University of Michigan who originally suggested the onocleoid ferns as a possible research problem, and who provided many helpful suggestions. Funds for field work in Mexico and Guatemala were provided by the Associates in Tropical Biogeography of the University of California. Funds for travel to the eastern United States for field studies of *Onoclea sensibilis* and *Matteuccia struthiopteris* were provided by the Research-Patent Fund of the University of California. Portions of the work reported here were done during the tenure of a University of California Graduate Fellowship. Appreciation is also expressed to the University of Hawaii for providing funds and facilities for some of the research.

Field collections were made by various people, and living material was sent to me for cultivation. I would particularly like to thank Dr. William Anderson of the University of Michigan for supplying material of *Onocleopsis hintonii* from southern Mexico and Dr. Y. Nozu, Botanical Institute, University of Tokyo, for sending plants of *Onoclea sensibilis, Matteuccia struthiopteris,* and *M. orientalis.*

During my field work in Mexico I was aided by several people. In particular, I would like to acknowledge the help of Dr. Eizi Matuda of the University of Mexico.

I would also like to thank Drs. Edward Klekowski, Jr., Robert W. Cruden, Richard S. Mitchell, and A. S. Foster, for their interest and helpful suggestions.

Lastly, I would like to thank the staffs of the herbaria that loaned materials for study at the University of California. These are listed below, preceded by the proper abbreviations from Lanjouw and Stafleu (1964):

BG	Universitetets Botaniske Museum, Bergen
BM	British Museum (Natural History)
E	Royal Botanic Garden, Edinburgh
GH	Gray Herbarium
H	Botanical Museum, Helsinki
K	Royal Botanic Garden, Kew
KYO	Kyoto University
LD	Botanical Museum, Lund
LE	Komarov Botanical Institute, Leningrad
LWG	National Botanic Garden, Lucknow
MICH	University of Michigan
MO	Missouri Botanical Garden
NY	New York Botanical Garden
O	Botanisk Museum, Oslo

P	Museum National d'Histoire Naturelle, Paris
S	Naturhistoriska Riksmuseet, Stockholm
TAI	National Taiwan University
TI	Botanical Institute, University of Tokyo
TNS	National Science Museum, Tokyo
TROM	Tromsoe Museum
UC	University of California, Berkeley
UPS	University of Uppsala
US	United States National Herbarium

HISTORY OF INTERPRETATION OF THE ONOCLEOID FERNS

Although *Onoclea, Onocleopsis,* and *Matteuccia* have been variously placed taxonomically there has been a general consistency in grouping them within either the Aspidiaceae or the Blechnaceae. Moore (1857) placed the onocleoid genera in the subtribe Aspidieae of the Polypodiaceae with the genera *Aspidium, Cyrtomium, Polystichum, Faydenia, Lastrea, Oleandra,* and *Nephrolepis.* He remarked that *Onoclea* is "A very elegant and distinct genus, which we think Mr. Smith correctly refers to the Aspideae, though the nature of the special indusia is not easily made out" (p. lxxxi). W. J. Hooker (1874) placed *Onoclea* (including *Matteuccia*) in the suborder Polypodiaceae, tribe Dicksonieae, with *Woodsia, Dicksonia, Hypoderris, Peranema,* and *Athyrium.* He also separated Davalliaceae, Aspidiaceae, Blechnaceae, and Aspleniaceae into distinct tribes. He defined the Dicksonieae as including those ferns with (1) globose sori, situated on the back or apex of a vein, with a free, closed, inferior, subglobose involucre, sometimes covering the whole sorus, that at length bursts irregularly; or more frequently is cup-shaped, entire or bilabiate; (2) a caudex that is rarely arborescent; and (3) free or anastomosing venation in the fronds.

Diels placed the onocleoid ferns with the woodsioid genera (including *Diacalpe, Peranema, Woodsia, Hypoderris, Cystopteris,* and *Acrophorus*) although he stated that "Die Verwandtschaft dieser isolierten Subtribus mit der verigen ist ausserst zweifelhaft" (1902: 164).

Christensen (1906) followed Diels' classification in nearly every respect. In his later system (1938), however, he erected the subfamily Onocleoideae of the Polypodiaceae and remarked that it formed a link between Cyatheaceae and the blechnoid ferns.

Bower (1908) considered that *Onoclea* and *Woodsia* formed a phylogentic series, a phyletic line of progression from a basipetal to a mixed sorus traceable from *Onoclea* and *Diacalpe* to *Woodsia* and *Hypoderris.* He stated that the onocleoid sorus is characteristic of the Cyatheaceae but that the sporangium is similar to that of the Polypodiaceae *sensu lato.* Bower (1928) thought that there was obvious affinity between the onocleoids and the woodsioids, but that this reflects their common origin rather than any present close relationship, specifically, a common cyatheoid ancestry. The onocleoids were visualized as a synthetic group connecting the cyatheoids with the blechnoids. He cited the following differences of the onocleoids from the cyatheoids: (1) smaller size and simpler anatomy; (2) heterophylly; and (3) seasonal leaf fall. The first character is present in such cyatheoids as *Lophosoria,* whereas the second and third are essentially limited to the onocleoids and blechnoids. Bower stated: "The Onocleoid ferns may be regarded as a series of advance running parallel to that of the Woodsieae, and like them they have carried the more primitive Cyatheoid type into boreal and alpine climates. There additional protection was a necessity, which has here been achieved partly by their firm texture and by seasonal leaf-fall, partly by heterophylly, and partly by their investure with broad protective scales. The two parallel series may be held as outstanding colonists, in which the leaves more especially have been modified so as to meet the conditions of the more exacting habitats to which they have spread. They may in fact be regarded respectively as arctic and xerophytic derivatives from the more primitive Cyatheoids" (1928: 159–160).

Since Bower considered the onocleoid ferns to be a separate phyletic offshoot from the cyatheoids, he designated them a distinct tribe. He also stated that "From the time of Willdenow and other early writers a relation has been recognized between the onocleoid ferns and *Blechnum.* There is general similarity of habit, but an obvious objection would seem to be the presence of a true indusium in the former, and its absence in the latter. . . . Glandular hairs of the Blechnoid Type are present though sparingly on the young pinnae" (1928: 155). Bower also mentioned that the lack of an indusium in *Matteuccia intermedia* indicated a link to the blechnoid ferns. He stated that there were good grounds, both soral and anatomical, for comparing *M. intermedia* with the simpler blechnoid ferns. He envisaged a relationship downward of *Matteuccia* to *Alsophila* and *Lophosoria* and upward to the genus *Blechnum* and suggested that "the Blechnoids probably sprang from some early Cyatheoid source, and the nearest indication of a connection is seen in . . . *Matteuccia intermedia*" (1928: 157).

Ching (1940) regarded the onocleoids as an offshoot of a phyletic line running from cyatheoid stock through Woodsiaceae and Peranemaceae to Aspidiaceae. His arrangement does not differ materially from that of Bower.

Copeland stated that the collective affinity of the onocleoids "has been the subject of wide, not to say wild, speculation. Bower has overemphasized their resemblance to *Cyathea,* which, except for features common to the Dryopteroid ferns as a whole, extends only to the elevated receptacle and consequently inferior indusium" (1947: 104). Copeland thought many of the common characters a result of homoplasy and found no evidence of relationship. He reported that "The commoner association of these two genera with *Woodsia, Diacalpe,* and *Peranema* has more in its favor. Every item of resemblance to *Cyathea* recurs here, plus similarity of sporangia and spores. There is hardly ground for doubt that in this case the affinity is real" (1947: 104).

Alston (1956) was one of the first to utilize presence or absence of the perispore in a classification system. In his paper on the subdivisions of the Polypodiaceae *sensu lato,* he erected six families, including Aspleniaceae, Thelypteridaceae, Athyriaceae, Aspidiaceae, Lomariopsidaceae, and Blechnaceae. *Matteuccia* was placed in the family Athyriaceae with *Athyrium, Diplazium, Woodsia, Stenolepia, Cheilanthopsis,* and *Cystopteris.*

Pichi-Sermolli (1958) endorsed Alston's classification and stated that it agreed better than did any previous system with his concept of the phylogeny of ferns.

Holttum (1947) placed many of the more advanced ferns in the large family Dennstaedtiaceae and included in this family eleven subfamilies. Dennstaedtiaceae included such subfamilies as Blechnoideae, Pteridoideae, Oleandroideae, Davallioideae, Athyriodeae, and Dryopteridoideae. Concerning the onocleoids, Holttum commented "I do not accept this as a group intermediate between Cyatheaceae and *Blechnum.* I doubt whether it is nearly related either to Cyatheaceae or to Dicksoniaceae, and leave it at present unplaced, pending further investigation" (1947: 124). In 1965, he added that the genus *Onoclea* was usually associated with *Matteuccia* but "has important differences including chromosome number, and its place seems uncertain" (p. 209).

Morton (1965, personal communication) stated that *Dryopteris, Athyrium,* and *Blechnum* are the obvious allies of the onocleoids, and that the contraction of the fertile frond is a result of homoplasy rather than an indicator of relationship.

Wagner (1969) has recently developed a phylogenetic scheme for the Pteridophyta which is quite similar to that proposed by Holttum (fig. 1). In his large group Indusiatae he includes nearly all the families Holttum considered as part of Dennstaedtiaceae. He believes that Aspidiaceae, Aspleniaceae, Blechnaceae, and Davalliaceae are very closely related and have evolved from dennstaedtioid ancestors. He does not deal with the onocleoids in his system.

Onocleopsis has only recently been published and has not been considered in any of the above treatments.

There appears to be general agreement among the authors cited above that the onocleoid ferns are part of the great phyletic line stemming from the dennstaedtioid ferns and culminating in Aspidiaceae *sensu lato,* including Blechnaceae, Davalliaceae, and Aspleniaceae, and it is with this group that the complex has been most closely associated. Within this group, the principal genera with which the onocleoids have been associated are *Athyrium, Blechnum, Cystopteris, Diacalpe, Diplazium, Dryopteris, Faydenia, Lastrea, Peranema, Polystichum,* and *Woodsia.* All of these genera are members of either Blechnaceae or Aspidiaceae *sensu* Wagner and Copeland. This study has drawn evidence primarily from these genera in hopes of obtaining information that may shed some light on the actual affinity of the onocleoid ferns. It is clear that a detailed anatomical and morphological investigation of these groups will provide a firmer basis for comparison than the utilization of only a few important characters determined a priori, which has been the common practice of many pteridologists in the past. Until now, no critical comparative study has been made either of the onocleoid ferns themselves or of their relation with any of the above groups.

COMPARATIVE MORPHOLOGY

Methods and Materials

The morphology of the onocleoid ferns and the purportedly related genera within Aspidiaceae and Blechnaceae was analyzed comparatively following the fern life-cycle. This was done in an attempt to document the relative distribution within these groups of the characteristics found in the onocleoid complex. It is hoped that in this way the data compiled from a complete analysis of life-cycles will indicate probable affinity of the onocleoid ferns.

The data on which this study is based have been obtained either from plants cultivated at the University of California Botanical Garden or from herbarium sheets. Specimens that document this study are listed in the Appendix.

Anatomical material was killed and fixed in F.A.A. (50 percent ETOH), sectioned at 12 to 15 microns, and stained according to Northen's modification of Foster's technique. Techniques for the handling of other material will be discussed below under the specific heading for the feature concerned.

Stomata measurements below refer to the size and shape of both guard cells and the enclosed pore.

Spores

Methods.—Spores were taken from both living plants and herbarium sheets and mounted in diaphane. In most cases the perispore was thin enough to allow measurement of the exospore directly without prior removal of the perispore. Spore size was constant in both living and dried material.

Onoclea sensibilis **L.**—Spores of *O. sensibilis* are bilateral, monolete, slightly concave-convex in lateral view, ovate in polar view, and bear a prominent perispore (perine). The perispore is loosely attached and is sometimes disjoined easily. It is nearly hyaline to brownish in color, granulose, and sparsely folded into small ridges. The perispore protrudes 2 to 7 (**x*** = 4) microns from the outer surface of the exine. The exine is about 2 microns thick, hyaline, and smooth. Spores of *Onoclea* without the attached perispore are 47 to 85 (**x** = 62) microns long and 32 to 68 (**x** = 42) microns wide. (Table 2 summarizes all measurement of spores.) They contain numerous green chloroplasts 2 to 5 microns in diameter. As the spores deteriorate following maturity the plastids nearest the exine gradually lose their color and become hyaline.

Sectioning of mature spores of *O. sensibilis* has proven to be extremely difficult. Marengo (1949, 1956) found that spores were penetrable only by the strongest fixatives and that sectioning difficulties were caused primarily by the tough, brittle nature of the inner spore wall. He suggested that the inner spore wall was impregnated with silica, a hypothesis borne out by the facts that concentrated H_2SO_4 would not dissolve the material and that x-ray diffraction produced consistent quartz lines following incineration. Upon sectioning the spore he found a centrally placed nucleus with a well-defined nuclear membrane and nucleolus. The chloroplasts were densely crowded around the nucleus, and the periphery of the spore contained a mass of deeply staining granules. The latter he interpreted as either immature plastids or granules of ergastic material.

Matteuccia.—Spores of all species of *Matteuccia* are bilateral, monolete, slightly concave-convex in lateral view, ovate in polar view, and have a prominent perispore. The perispore is nearly hyaline to brown, folded into prominent, irregular thick ridges, and more or less easily disjoined from the exospore. In *M. orientalis* the perispore is very thin and extends up to 23 microns from the exospore. The perispore of *M. struthiopteris* is light brown, more strongly winged than that of

* **x** = mean.

M. orientalis, and extends up to 18 microns from the exospore. It is granular in appearance and is easily detached by applying pressure to the cover slip.

The exospore of all species of *Matteuccia* is smooth, without ornamentation, and 2.0 to 2.2 microns in thickness. Nayar and Devi (1964) reported the laesura to be tenimarginate and about 42 microns long. Chloroplasts in spores of *M. struthiopteris* are about 2.0 to 2.5 microns in diameter. In older spores the plastids nearer the exospore surface become senescent first, losing their color but maintaining their shape.

Spores of *M. struthiopteris* without perispore are 40 to 55 (**x** = 48) microns long and 24 to 40 (**x** = 34) microns wide. The spores of *M. orientalis* exhibit two size classes correlated with geographic source which, in turn, probably reflects ploidy level. Spores of mainland Asiatic plants are 57 to 85 (**x** = 70) microns long and 30 to 60 (**x** = 48) microns wide, without perispore. Spores of Japanese plants are considerably larger with the length varying from 65 to 95 (**x** = 79) microns and the width from 37.5 to 70 (**x** = 53) microns. This correlation with geography holds true for all plants studied except for one collection from southeast China (*Rock 7569,* Yunnan, UC). Spores of *M. intermedia* are 60 to 80 (**x** = 68) microns long without perispore.

Marengo (1956) found that spores of *M. struthiopteris* were slightly smaller than but otherwise identical with those of *Onoclea sensibilis.* The perispore was easily removed and the exospore was of the same thickness except that it was easily sectioned, implying a lack of silica.

Onocleopsis hintonii Ballard.—The spores of *O. hintonii* are similar in nearly all respects to those of *Onoclea* and *Matteuccia.* They are bilateral, monolete, slightly concave-convex in lateral view and ovate in polar view, bearing a prominent perispore. The perispore is yellow-brown to brown and more or less loosely attached. It extends from 5 to 20 microns from the exospore and is generally folded into prominent irregular ridges. The exospore is smooth, without ornamentation, and about 2 microns thick. The perispore of some spores within the sporangium appears to be a thin wing about 7.5 microns thick extending uniformly around the exospore. The spores are green because of numerous chloroplasts scattered uniformly throughout the cell, although the color is not readily apparent because of the thickness of the perispore. The spores are 45 to 53 (**x** = 50) microns long and 32 to 35 (**x** = 34) wide without perispore.

Perispore.—The perispore has proven to have considerable taxonomic value. Bower stated: "The spore-wall is often marked by characteristic sculpturing. But more important than this is the presence or absence of a deposit of the perispore from the tapetum" (1923:259). He noted that a perispore was found neither in the eusporangiate ferns, nor in Osmundaceae, Schizaeaceae, Gleicheniaceae, Hymenophyllaceae, Cyatheaceae, Davallieae or in *Ceratopteris.* Among the more advanced ferns he found it lacking in the Vittariae, Gymnogrammeae, Polypodieae, and Pterideae. He did report its occurrence in Aspleniae and Aspideae. Bower considered it to be a recent feature apparently restricted to certain circles of affinity. The perispore was employed in natural classification as long ago as 1911 by Hannig. That author stated that this character is generally important in grouping genera within Polypodieae, Aspleniae, and Aspideae, and that the polypodioid spore is characterized by the absence of a perispore and the aspidioid spore is characterized by its presence. Consistent use of this structure in taxonomy of ferns was first made by A. H. G. Alston (1956) in his short paper on the subdivisions of Polypodiaceae found in West Tropical Africa. He used the perispore in combination with spore-type and the occurrence of an indusium to segregate distinct families from the Polypodiaceae *sensu lato.*

The perispore is lacking in more primitive ferns and in nearly all ferns with trilete spores. Tschudy and Tschudy (1965) found a perispore or a perispore-like structure in *Cyathea, Anemia, Lygodium, Hemionitis, Lonchitus,* and *Adiantum,*

TABLE 1

CHARACTERISTICS OF CHLOROPHYLLOUS SPORES

Taxon	Spore shape	Perispore
Equisetaceae	Spherical	Absent
Osmundaceae	Tetrahedral	Absent
Grammitidaceae	Tetrahedral	Absent
Hymenophyllaceae	Tetrahedral	Absent
Polypodiaceae		
Christiopteris	Bilateral	Absent
Marginariopsis	Bilateral	Absent
Aspidiaceae		
Lomariopsis	Bilateral	Present
Onocleoideae	Bilateral	Present

all genera with trilete spores. In many cases the perispore was visible only after treatment by acetylation and chlorination. Concerning the perispore of *Anemia* they remarked, "Whether or not this is a true exospore we are not at present prepared to say. It, however, exhibits all of the characteristic features of an exospore" (p. 17).

Generally the perispore can serve as a valid diagnostic character especially at the family level. The saccate or winged perispore is almost entirely restricted to the Aspidiaceae, Blechnaceae, Davalliaceae, and Aspleniaceae. Outside of this group a winged perispore is known only in *Polypodium chnoodes* Spreng., a common species of Central America (Lloyd, 1969).

I have investigated the perispore in order to determine its basic overall morphology, type and configuration of wing structure, and size. However, careful investigation of all perispore-bearing species is required to determine the constancy of this particular character in phylogeny.

Occurrence of chlorophyllous spores.—Spores containing chloroplasts are known to occur in several unrelated families of pteridophytes including: (1) Equisetaceae—all species; (2) Osmundaceae—all species; (3) Grammitidaceae—all species; (4) Polypodiaceae—*Christiopteris* and *Marginariopsis;* (5) Hymenophyllaceae—all species; (6) Blechnaceae—*Blechnum nudum* (Stone, 1961); (7) Aspidiaceae—*Lomariopsis sorbifolia* (Tschudy and Tschudy, 1965); (8) onocleoid ferns—all species. Table 1 summarizes the spore characteristics of most of these species.

It is very difficult to determine spore color from any but the freshest herbarium specimens. Spores of *Osmunda,* for example, will begin to lose their green color after several weeks in storage. Spores of the onocleoid ferns lose their color in about one year. A bleaching of the plastids near the margin of the exospore progresses gradually toward the center of the spore. Spores of *Marginariopsis,* a genus of disputed relationship between Grammitidaceae and Polypodiaceae, contain only a few chloroplasts when compared with the larger numbers present in other ferns. These break down almost immediately after the plants are collected and the spores are bright yellow after only one or two days. Nayar (1967) reported the spores of *Christiopteris* as deep yellow with a greenish tinge, the latter being the effect of many small pale green plastids.

The occurrence of chlorophyllous spores in ferns is so rare that it could be used as a good indicator of phylogenetic relationships if correlated with other morphological characteristics. This is true for such isolated groups as Osmundaceae, Equisetaceae, Grammitidaceae, and Hymenophyllaceae. The Osmundaceae and the Equisetaceae are two independent isolated evolutionary lines which have been distinct for a long time, perhaps since the Carboniferous. The Hymenophyllaceae and the Grammitidaceae are also monophyletic lines stemming from dennstaedtioid and polypodioid ancestors, respectively (Wagner, 1969). All species of these four groups have green spores, and it is reasonable to assume that this characteristic is a relatively ancient one.

The occurrence of chlorophyllous spores in the more advanced groups of Aspidiaceae, Blechnaceae, and Polypodiaceae must be interpreted differently, however. Although it is possible that this character is relatively ancient in these groups, its absence in closely related species as well as from most of the species of the aspidioid-group indicates a high probability that it is of more recent derivation. This is apparently the case in *Christiopteris* and *Marginariopsis* (Polypodiaceae), *Lomariopsis* (Aspidiaceae), and *Blechnum* (Blechnaceae). This might also be the case as far as the onocleoid genera are concerned except that this group is held together also by a series of other morphological features. In this case it can be considered to be one more feature that supports common ancestry for the group.

Comparisons with other genera.—Spores of all genera under consideration as possible relatives of the onocleoid ferns were examined carefully. It was concluded that the only characteristic that might be indicative of phyletic relationship is the type and form of perispore. Generally, most species within Blechnaceae have a reddish-brown uniform perispore or a membranous hyaline outer perispore with a thicker reddish-brown base. The perispore of Blechnaceae may or may not have folded ridges running from the inner to the outer ridges (Nayar et al., 1966:408, figs. 1–8). Of all species of Blechnaceae investigated only *Woodwardia areolata* has a structure approaching that of the onocleoid ferns. In *W. areolata* the perispore is light brown to hyaline in color, 7 to 18 microns thick, granular, and sparsely folded into small ridges. The structure appears to be, however, much more membranous than the onocleoid perispore.

Within the Aspidiaceae all perispores observed showed significant differences with those of the onocleoid ferns, except for that of *Athyrium nakanoi.* The perispore of the latter species is very similar to that of *Woodwardia areolata* and differs from the onocleoid fern structure in the same characteristics.

Conclusion.—The spores of the onocleoid ferns show a much greater affinity in structure with themselves than with any other species within either the Blechnaceae or Aspidiaceae. The Blechnaceae as a whole appear to exhibit great consistency in the structure of the perispore, except for that found in *Woodwardia areolata.* As the onocleoid fern spores show none of the typical blechnoid perispore characteristics, it is apparent that there is no close relation here. On the contrary, the relative uniqueness of the onocleoid perispore fits well into the diverse patterns as presented by species of the Aspidiaceae, although there is no direct correlation indicated with any of the groups studied. Table 2 summarizes all of the measurements made for all species investigated.

Gametophyte

Methods.—Gametophytes were studied entirely from agar cultures because visits to natural habitats failed to yield sufficient numbers for study. Culturing also standardized conditions under which gametophytes were grown. Spores were sown and gametophytes cultured on an inorganic nutrient medium solidified with 1 percent agar (see Klekowski, 1969, for composition of medium). The cultures were grown under continuous fluorescent illumination at an intensity of 210 to 290 ft-c. For morphological studies, gametophytes were examined either fresh or were mounted in Hoyer's Medium (Beeks, 1955) and stained with aceto-carmine. Other gametophytes were stained with sudan IV for elucidation of the chemical nature of deposits on cell surfaces as well as exudates from the prothallial hairs. Table 3 summarizes the gametophytic characteristics of onocleoid ferns.

Onoclea sensibilis.—Spores of *Onoclea* germinate from one to three days after sowing (rarely within 13 hours of sowing). The rhizoid emerges first in most cases through the pore of the swollen spore (fig. 2, *a*). This is followed by emergence and proliferation of the first prothallial cell (fig. 2, *b*). Chloroplasts in the first prothallial cell accumulate near the apical end preparatory to the first transverse division of the filament. After six days the prothallial filament is composed of three barrel-shaped cells 170 to 230 microns long (fig. 2, *c*). In one instance the second filament cell had divided periclinally, initiating two-dimensional growth in the gametophyte. In most cases, the filament reaches a four- or five-celled stage before the terminal or subterminal cell divided periclinally (fig. 2, *e, f*). After fifteen days, gametophytes had attained a thirty-cell stage with a definite apical cell and the first appearance of the initiation of the wings of the prothallus. There is some differentiation in growth rate apparent as other gametophytes in the same culture are still in the one-dimensional stage with three to five filament cells. Twenty days after sowing, the first prothallial trichomes appear near the apex of the gametophyte on one or both of the lateral wings (fig. 2, *j*). At this stage the gametophyte has attained a length of 400 microns. The prothallial trichomes at this age do not produce any glandular secretions, and are unicellular and unbranched. Some of the trichomes contain a few, minute chloroplasts (fig. 2, *p*).

By the twenty-fifth day unicellular trichomes have appeared along the lateral margins of the now cordate gametophyte, which is about 700 microns long and 400 microns wide (fig. 2, *n*). The spore wall may still be attached to the first filament cell, and rhizoids have begun to proliferate from the second and third filament cells.

TABLE 2

Spore Characteristics and Annulus Number of Selected Genera of Aspidiaceae and Blechnaceae

Genus	No. ind. ann. cell		Spore length				Spore width				No. species investigated
			w/p		w/o per.		w/p		w/o per.		
	Range	Mean	Range	Mean	Range	Mean	Range	Mean	Range	Mean	
ASPIDIACEAE											
Athyrium	11–18	14	30–65	52	28–50	40	20–48	36	17–35	25	5
Cystopteris	13–15	14	45–52	49	42–48	44	30–38	35	25–33	30	1
Diacalpe	12–16	14	45–57	53	40–45	43	25–35	30	17–28	22	1
Diplazium	12–18	15	42–75	52	30–50	38	27–50	38	20–33	27	5
Dryopteris	13–19	15	37–55	49	.....	..	25–45	36	....	..	3
Faydenia	12–17	14	60–75	69	50–58	52	50–58	53	27–43	38	2
Lastrea	12–18	14	40–50	45	30–48	37	25–33	29	20–30	24	3
Peranema	13–15	14	40–53	46	32–38	36	27–40	33	20–28	23	2
Polystichum	12–19	15	37–43	40	32–35	34	27–40	32	22–28	25	2
Woodsia	10–29	17	47–60	51	40–50	46	32–48	39	27–38	33	4
BLECHNACEAE											
Blechnum	12–24	16	37–80	56	35–68	48	25–60	39	20–48	32	6
Brainea	14–20	18	40–55	49	35–50	45	30–45	40	25–40	35	1
Doodia	11–22	15	27–53	38	25–50	35	20–35	27	17–33	24	3
Sadleria	13–24	18	47–103	69	45–80	58	32–75	47	30–50	38	3
Woodwardia	19–27	22	45–75	60	37–53	46	37–58	46	20–38	34	1
ONOCLEOIDS											
Onoclea	22–41	28	52–90	69	47–85	62	37–75	47	32–68	42	1
Onocleopsis	21–26	24	55–70	62	45–53	50	37–52	44	32–35	34	1
Matteuccia struthiop.	21–30	24	47–73	57	40–55	48	35–50	43	25–40	34	1
M. orientalis	27–45	35	72–110	89	57–95	74	42–80	61	30–70	50	1
M. intermedia	28–36	32	.....	..	60–80	68	.....	..	.....	..	1
ASPIDIACEAE	10–29	15	30–75	51	27–58	41	20–58	36	17–43	38	28
BLECHNACEAE	11–27	18	27–103	54	25–80	46	20–75	40	17–50	32	14
ONOCLEOIDEAE	21–45	28	47–110	69	40–95	58	35–80	48	25–70	40	5

All measurements in microns.
w/p = with perispore.
w/o per. = without perispore.

TABLE 3

SUMMARY OF GAMETOPHYTIC CHARACTERISTICS OF ONOCLEOID FERNS
(Measurements in microns)

Character	*Onoclea*	*M. struthiopteris*	*M. orientalis*	*Onocleopsis*
Shape	cordate, irreg.	cordate, irreg.	cordate	cordate, strap
Trichomes	present	absent	present	absent
Length	30–40	...	32–92	...
Archegonium-diameter	70	48–60	40–45	62–75
Direction neck orientation	basally	basally	basally, all	basally
Number neck cells	4–7	3–7	4–6	3–4
Number neck tiers	4	4	4	4
Archegonium-length	90–125	100–110	100–135	95–125
Antheridium-diameter	50–67	60–70	60–88	53–88
Number of filament cells before first periclinal division	4–5	4–5	4–5	(1) 3 (4)
Length of apical cell				
Young prothallus	32	28	30	25
Mature prothallus	52	...	50	47

Mature gametophytes of *Onoclea* initiate either archegonia or antheridia at the onset of sexual maturity (see Klekowski and Lloyd, 1968, for discussion of the reproductive biology of this species). Female gametophytes are about 9 mm wide and 2.5 mm long with a well-developed cushion on which are borne the archegonia. The apical region of the gametophyte is composed of 12 to 17 rectangular apical initials, each about 40 microns long, and 10 microns wide. The gametophyte is profusely covered with small unicellular, clavate trichomes, 30 to 44 microns long and 12 to 23 microns wide at the apex. At maturity these trichomes secrete a substance, interpreted as a fat or wax since it reacts positively to sudan IV stain (fig. 2, *q*). Cells near the margin of the gametophyte, as well as those of the wings, also secrete a fatty substance which reacts with sudan IV. The basal portion of the prothallus is densely covered with unbranched brownish rhizoids.

Archegonia are initiated near the apical region and mature gradually as the cushion elongates. At maturity they are about 125 microns long and 70 microns wide at their broadest point. The neck of the archegonium is composed of four to seven cells on the side toward the apical notch of the gametophyte and three to four cells on the basal side. The neck cells are four-tiered (fig. 2, *o*). There are chloroplasts within the neck cells but fewer than in the cells of the thallus. Internally, the mature archegonium consists of the basal egg cell, a ventral canal cell with an elongate nucleus, and two neck-canal nuclei (fig. 2, *l*). There is no apparent cell wall between the neck-canal nuclei. This condition was also reported by Wagner (1952) for *Diellia*.

The archegonia are borne on a cushion eight to fourteen cells thick; this is in contrast to the normal one-celled thickness of the wings of the gametophyte. The necks of the archegonia are reflexed away from the notch at an angle of 45° to 50° in response to the difference in number of neck cells present on the apical and basal sides. In my cultures, archegonia were borne on both sides of the gametophyte although they usually appeared on the lower side. This is probably

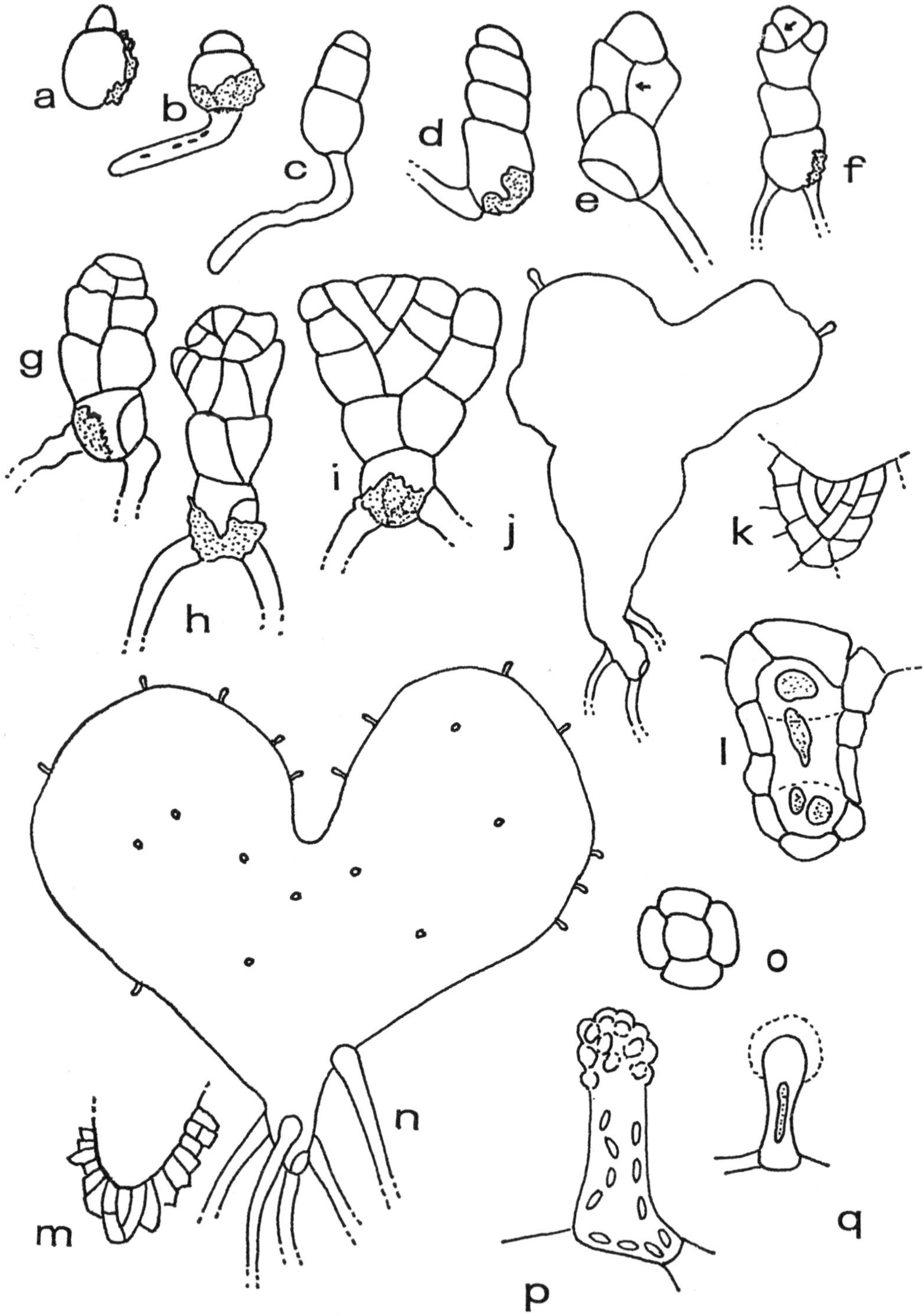

Fig. 2. Ontogeny of the gametophyte of *Onoclea sensibilis*. *a*, spore germination, × 188; *b*, first prothallial cell, × 188; *c*, 3-prothallial cell stage, × 125; *d*, 4-prothallial cell stage, × 135; *e*, initiation of two-dimensional growth by division of penultimate cell, × 170; *f*, initiation of

in response to the stimulus of light striking the gametophyte on all sides. Rarely a neck cell of an archegonium would bear a typical gametophytic clavate trichome.

Older cordate female gametophtyes will usually proliferate new tissue off the wings, which will bear antheridia. Female gametophytes which do not produce sexual sporophytes will elongate and rarely produce an apogmous outgrowth.

Male gametophytes of *Onoclea* vary greatly in size and shape, depending primarily on the density of sowing. When mature, that is, when sex organs are initiated, the gametophytes may vary in shape from cordate, about 3 mm wide, to more or less spatulate or oblong, much reduced in size. The typically cordate male gametophyte commonly bears antheridia on both the cushion region and on the inner two-thirds of the wing surface. The apical growing region consists of five to six apical initials, each about 50 microns long and 10 microns wide. The margins and surfaces of the gametophyte are covered with numerous clavate trichomes, similar to those of the female gametophyte, but slightly smaller. The mature antheridium is of the most advanced type reported for leptosporangiate ferns. It is dome-shaped, with a basal ring cell, a second ring cell, and a terminal cap cell. These observations essentially agree with those of Campbell (1905). The diameter of the mature antheridium varies from 50 to 67 microns. Staining by sudan IV indicates the presence of waxy deposits on the surface of the antheridium, especially near the margins of the cap cell.

Male gametophytes that bear a cushion will in time produce archegonia and become hermaphroditic. Gametophytes of irregular size and shape and without a cushion have not yet been seen to produce archegonia and perhaps do not do so under any conditions.

Matteuccia.—The ontogenies of the gametophytes of both *Matteuccia struthiopteris* and *M. orientalis* (tetraploid) are very similar to that of *Onoclea* and will not be described in detail here (see figs. 3 and 4).

Prothalli of *M. struthiopteris* are regularly dioecious, some producing a cushion and archegonial meristem, and others remaining one-layered, of cordate or irregular form, and producing only antheridia. Many of the archegoniate prothalli will, during later development, produce antheridia on the lower portion of the wings, generally in the immediate vicinity of the rhizoids. Certain of the female gametophytes become hermaphroditic by proliferation from the margin of cordate tissue bearing antheridia.

Sexually mature male prothalli of *M. struthiopteris* are usually less than 2 mm in diameter, cordate or irregularly shaped, but usually without a cushion. A few of them will become hermaphroditic with age. Antheridia are borne on the lower two-thirds of the thallus. The mature antheridium is identical in structure to that of *Onoclea* and varies from 50 to 67 microns in diameter.

Female or hermaphroditic gametophytes are cordate in shape and usually larger than 2 mm in diameter. The archegonia, which are borne on the thickened cushion region, vary in structure. Archegonia from gametophytes of Japanese origin have short, four-tiered necks bearing four cells on the apical side and three

---

two-dimensional growth by division of terminal cell (arrow), × 135; *g, h, i,* further two-dimensional growth, × 147; *j,* cordate gametophyte with appearance of trichomes, × 147; *k,* apical cell, × 160; *l,* mature archegonium, × 280; *m,* apical region of mature gametophyte, × 160; *n,* young cordate gametophyte, × 70; *o,* x-sec. archegonium, × 215; *p,* trichome, × 1100; *q,* mature trichome with glandular secretion, × 450.

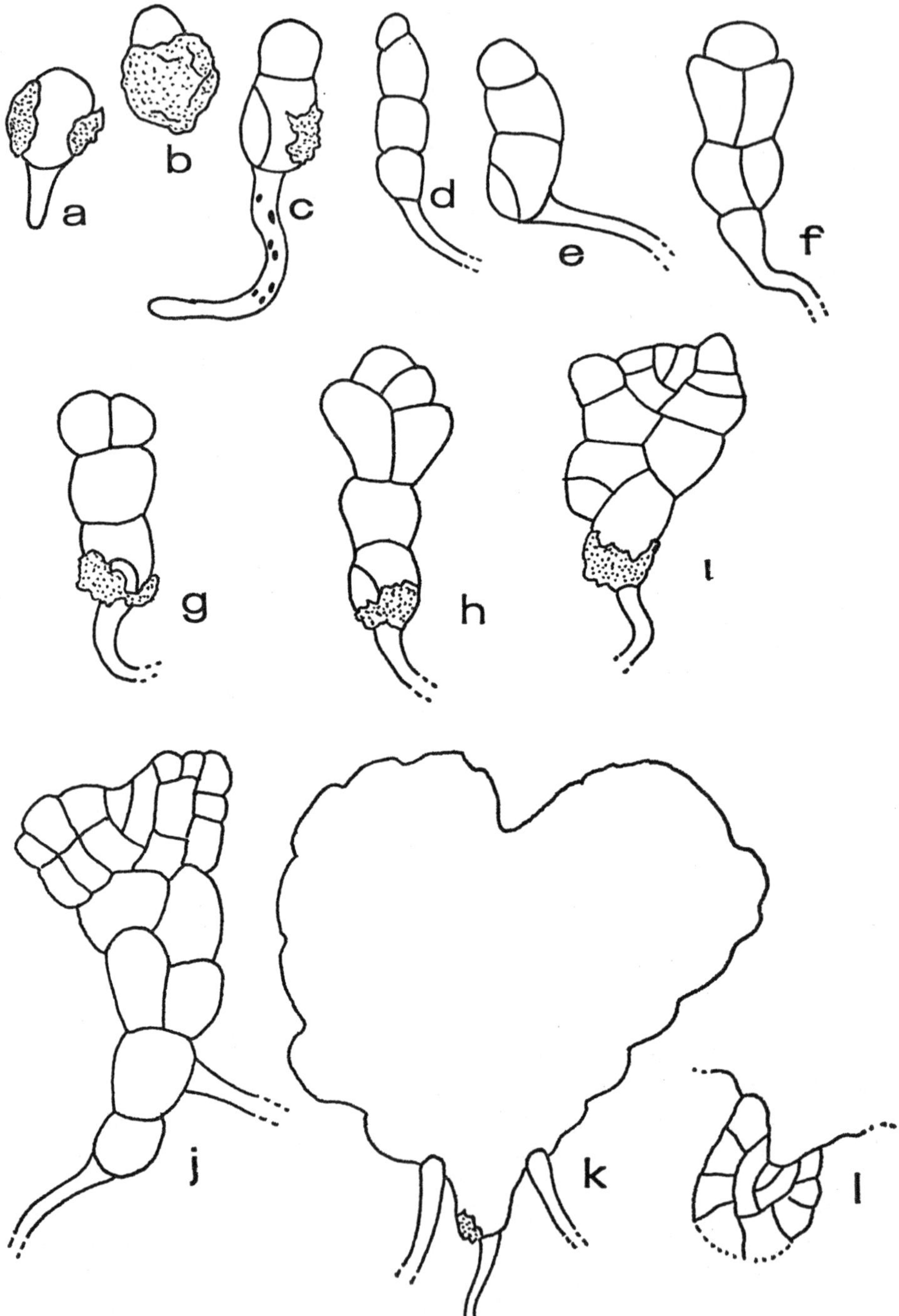

Fig. 3. Ontogeny of the gametophyte of *Matteuccia struthiopteris*. *a, b*, spore germination, × 250; *c*, one-dimensional filament, × 210; *d*, × 150; *e*, × 200; *f*, initiation of two-dimensional growth, × 175; *g*, initiation of two-dimensional growth by terminal cell-division, × 190; *h*, × 250; *i, j*, two-dimensional growth, × 160; *k*, young cordate gametophyte, × 180; *l*, apical cell of young cordate gametophyte, × 125.

on the basal side. This is in contrast to the observations of Momose (1958), who depicted archegonia with six and four cells, respectively, for plants of Japanese origin. Observations on plants from the eastern United States show archegonia identical in structure with those of *Onoclea,* bearing four to five (seven) cells on the apical side and three to five cells on the basal side. Some necks contain only five cells on both sides, those nearer the notch being the larger. The necks of *M. struthiopteris* bend basally at an angle of about 35°, and are about 90 microns long and 58 microns wide. The gametophytes of *M. struthiopteris* have been described in detail by Nayar (1968).

Older cordate gametophytes will elongate and rarely produce apogamous outgrowths.

Gametophytes of *Matteuccia orientalis* are either male or female and hermaphroditic, depending in the latter case upon age. Archegonia usually first appear on hermaphroditic prothalli followed by production of antheridia on the cushion. Rarely appearance of antheridia slightly precedes that of archegonia.

Male gametophytes are 1.7 to 3.3 mm in diameter, cordate, and bear antheridia on the basal one-third of the thallus. The apical meristem consists of about eight rectangular cells, each with a spindle-shaped nucleus. These cells are 25 to 32 microns long and 2.5 to 10 microns wide. Antheridia are 60 to 88 microns in diameter and are similar in structure to most of those of the higher leptosporangiate ferns.

Female and hermaphroditic gametophytes vary in size from 2.8 to 10 mm when sexually mature. The cordate gametophytes bear archegonia, or archegonia and antheridia in close proximity on the cushion. In no instance were antheridia observed on the wings of the prothallus. The archegonia have four-tiered necks, and are 100 to 135 microns long. The necks are broadly clavate with neck canal cells 40 to 45 microns in diameter. The egg cell at the base of the archegonium is 25 to 30 microns in diameter. The neck of the archegonium consists of five to six neck-canal cells. In some gametophtyes these appear to be extremely thin and primarily nuclear in content, with only a thin layer of surrounding cytoplasm. The archegonial necks of *M. orientalis* are oriented in all directions, although generally toward the base of the gametophyte.

Prothallial trichomes of Matteuccia.—Although the gametophytes of *M. struthiopteris* are glabrous, those of *M. orientalis* are densely covered with trichomes, which in structure resemble those of *Onoclea.* These trichomes are clavate when mature and vary from 32 to 92 microns in length (fig. 4, *k*). They are either one- or two-celled with the basal cell slightly enlarged and bearing a complex cytoplasmic structure. When mature, the trichomes secrete a waxy substance which reacts positively with sudan IV.

Onocleopsis hintonii.—The ontogeny of gametophytes of *Onocleopsis* (fig. 5) differs somewhat from that of both *Onoclea* and *Matteuccia.* However, there is reason to suspect that this may be caused in part by cultural conditions, especially light intensity. Gametophytes cultured at light intensities of 210 to 290 ft-c showed a marked tendency to proliferate into abnormal cabbage-like heads. Gametophytes cultured under lower light intensities showed a more nearly normal cordate shape when mature.

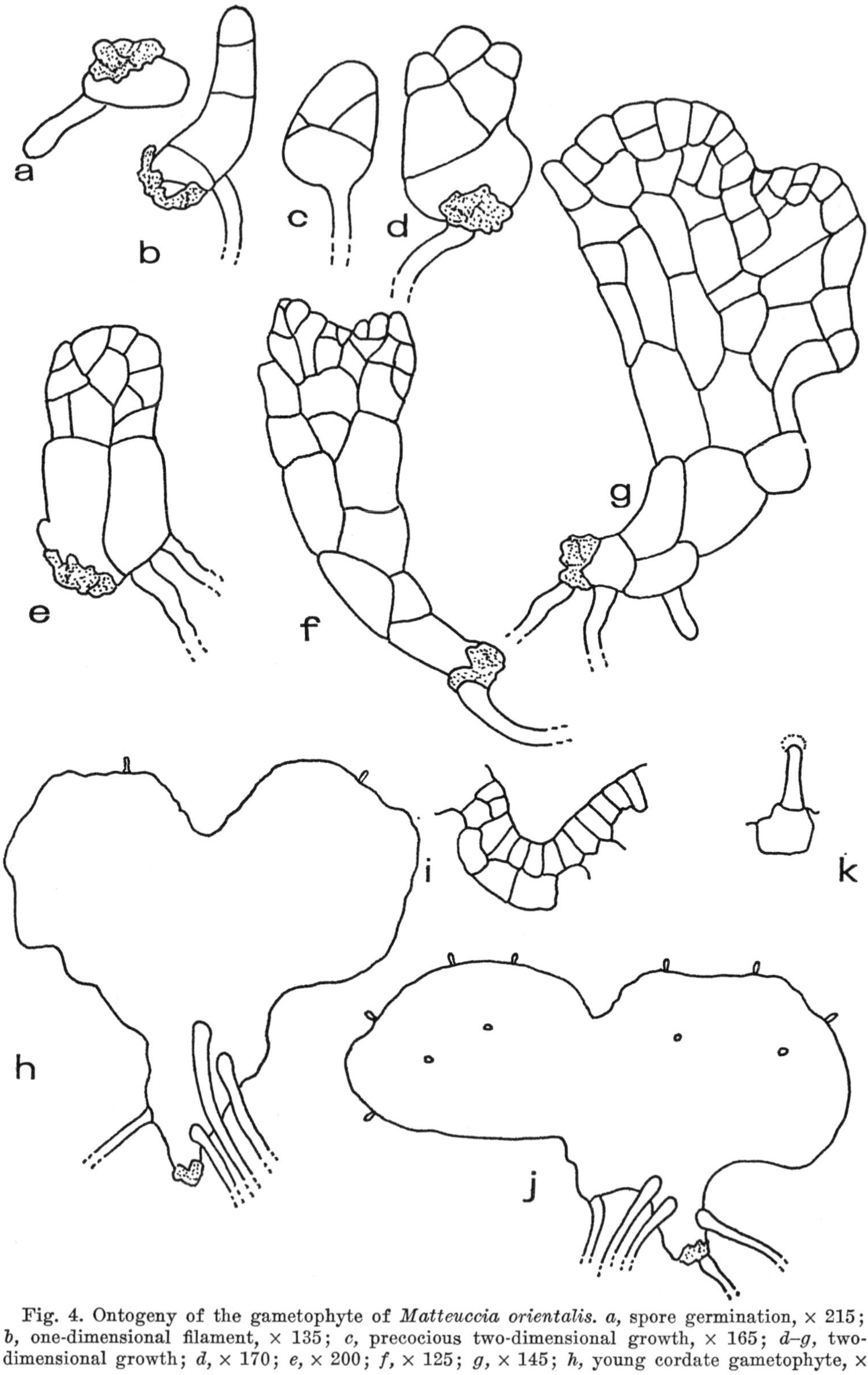

Fig. 4. Ontogeny of the gametophyte of *Matteuccia orientalis*. *a*, spore germination, × 215; *b*, one-dimensional filament, × 135; *c*, precocious two-dimensional growth, × 165; *d–g*, two-dimensional growth; *d*, × 170; *e*, × 200; *f*, × 125; *g*, × 145; *h*, young cordate gametophyte, ×

Gametophytes of *Onocleopsis* are extremely slow-growing when compared with those of *Onoclea* and *Matteuccia.* Spore germination is initiated either with the production of the rhizoid or the first prothallial cell (fig. 5, *a*). The young rhizoid is yellowish and contains a few chloroplasts. Subsequent growth of the young gametophyte will usually result in a one-dimensional filament of three or four barrel-shaped cells before periclinal divisions take place (fig. 5, *d*). These divisions have been observed to occur in the first prothallial cell, although they are more common in the third or fourth cell (fig. 5, *e, g*). In some gametophytes, each of the first three or four filament cells will cut off a lateral cell, giving the appearance of a gametophyte with three or four apical cells (fig. 5, *f*). By the time the length of the gametophyte has reached five cells, three or four periclinal divisions have occurred in the preceding cells. This produces a rather marked precocious two-dimensional form. Subsequent growth tends to produce relatively broad prothalli. In most cases the maturing gametophyte will produce an apical meristem and ultimately a cordate shape. However, growth is so irregular in pattern that some prothalli appear to have developed a lateral meristem (fig. 5, *m*). In many of my cultures the mature gametophyte was more nearly strap-shaped with two enlarged wings than it was cordate.

Unlike *Onoclea* and *Matteuccia,* the antheridia appear before the archegonia on gametophytes of *Onocleopsis.* These are borne on the wings and near the basal end of the prothallus. The mature antheridium is about 63 (53 to 65) microns in diameter and is similar in structure to those of *Onoclea* and *Matteuccia.*

Archegonia are initiated on the cushion about five cells behind the apical notch. The zone of archegonia extends basally in a narrow band on the cushion. The basally pointing neck of the mature archegonium varies from 85 to 100 microns in length and consists of three or four cells borne in four tiers. The internal structure of the archegonium is similar to that of *Onoclea.*

Gametophytes of *Onocleopsis* are devoid of trichomes. They react negatively to sudan IV stain except for a thin waxy covering on the outermost cells of the wings.

It is possible to correlate certain features of the gametophyte of *Onocleopsis* with the environment in which the mature sporophyte is found. Lack of trichomes, intolerance to light intensities used in culture, and lack of waxy cuticular deposits on the epidermal cells are possibly adaptations of the gametophyte to growth in the very mesic, dark environments in which this species is found.

Discussion of onocleoid fern gametophytes.—It is at once apparent that there are not yet enough detailed data available on gametophyte morphology to draw any definitive conclusions as to the phyletic importance of specific morphological characteristics. Our current knowledge limits our use of gametophytic morphology in fixing relationships only to the larger categories and phylogenetic lines. In closely related families, such as Aspidiaceae and Blechnaceae, most gametophytic characteristics are essentially similar. At levels below this it is impossible at this time to find reliable differentiating characteristics. For example, presence or absence of prothallial trichomes varies even within one genus, such as *Blechnum* and *Matteuccia,* in species which are considered related on sporophytic characteristics.

100; *i,* apical meristem of cordate gametophyte, × 150; *j,* cordate gametophyte, × 67; *k,* trichome, × 200.

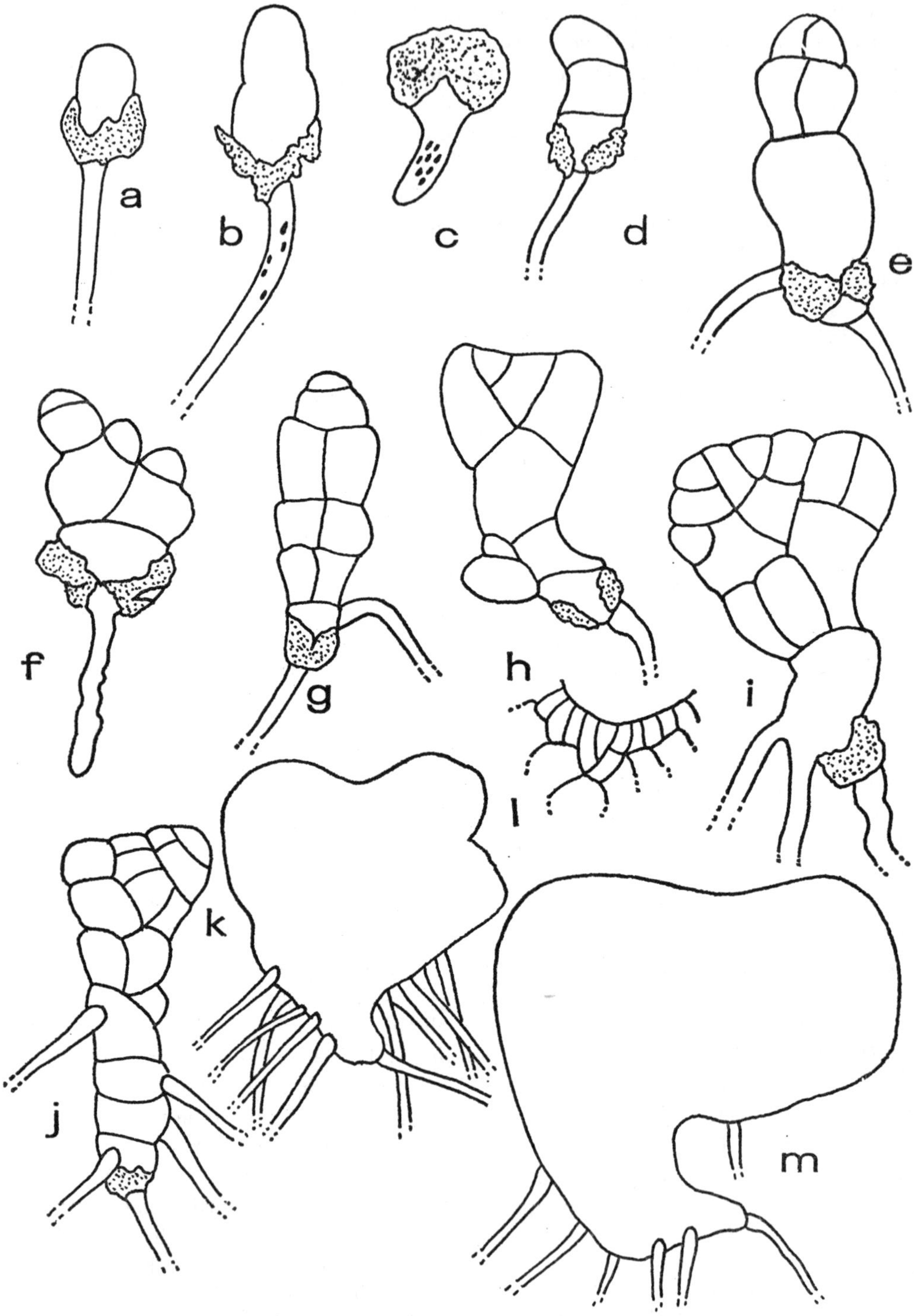

Fig. 5. Ontogeny of the gametophyte of *Onocleopsis hintonii*. *a*, spore germination, × 185; *b*, elongation of first prothallial cell, × 245; *c*, spore germination, × 550; *d*, 3-prothallial cell stage, × 145; *e*, initiation of two-dimensional growth, × 230; *f–i*, two-dimensional growth, ×

Recently, Momose (1958) has cited evidence from gametophytic morphology to support the generic distinction of *Pentarhizidium* (*P. orientalis* = *Matteuccia orientalis*) from *Matteuccia.* This evidence includes such features as presence of glandular trichomes, cellulose-notched margin, and location of antheridia on the prothallus. Nayar (1968) states that the developmental history of *M. orientalis* is different from that of *M. struthiopteris* and *Onoclea sensibilis,* although nowhere does he state how it differs.

Recent work by Klekowski (1968) and Klekowski and Lloyd (1968) has shown that the location and occurrence of antheridia and archegonia may be determined by the specialized nature of the breeding system rather than by phylogenetic origin alone. This also extends to the orientation of archegonial necks, either toward or away from the notch. It is apparent that developmental patterns can be altered drastically by the light intensities encountered during the culture of gametophytes. This was strikingly evident in the culture of *Onocleopsis* gametophytes, which produced proliferated, cabbage-head-like gametophytes at higher light intensities (e.g., those ideal for growth of *Onoclea* gametophytes) and more normal gametophytes at lower intensities (ca. 100 ft-c). In my studies I encountered no evidence that would support a generic separation of the species of *Matteuccia,* or for that matter, any evidence that would negate a relationship at the generic level among the onocleoid fern genera.

Comparisons with other genera.—Genera supposedly related to the onocleoid ferns were cultured from spore whenever possible. These included detailed studies of the following species: *Aspidium angulare, Athyrium* sp., *Blechnum gibbum, B. glandulosum, B. patersonii, B. procera, B. punctulatum, B. unilaterale, Cyrtomium falcatum, Diplazium cristatum, D. polypodioides, D. werkleanum, D.* sp., *Dryopteris aemula, D. setigera,* and *Tectaria cicutaria.* Additional data were compiled for other species from the abundant literature available on fern gametophytes (see Stokey, 1950, and Atkinson and Stokey, 1964, for a review of literature).

DEVELOPMENTAL PATTERN

The developmental pattern of the gametophyte of onocleoid ferns is similar in most respects to that found in aspidioid ferns. It is unlike that of most blechnoid ferns.

There appears to be a marked difference in developmental pattern between Aspidiaceae and Blechnaceae. This is apparently due to the appearance of a trichome on the terminal cell of the germ filament before formation of the plate. Nayar et al. (1966) report that in most species of *Blechnum* studied the germ filaments end in a trichome when they are three to five cells in length. This is also true of *Stenochlaena palustris* and *Doodia aspera.* They also report that in *Doodia* spp., *Woodwardia* spp., and some species of *Blechnum* the terminal cell of the germ filament becomes quiescent. Both of these phenomena tend to produce a laterally developed meristematic cell (Nayar et al., 1966, figs. 37, 50–62, 87). The lateral meristematic cell is also found in some of those aspidioid genera which produce trichomes on the terminal filament cell (Nayar and Kaur, 1964,

175; *j*, × 120; *k*, early cordate stage, × 80; *l*, apical cell of cordate gametophyte, × 135; *m*, young cordate gametophyte, × 95.

figs. 16–19, *Tectaria variolosa;* Nayar and Kaur, 1966, figs. 1-C, *Lithostegia foeniculacea;* and Lloyd on *"Aspidium" angulare* and *Dryopteris aemula*).

In most aspidioid genera studied, including *Acrophorus, Athyrium, Cornopteris, Ctenitis, Cyclosorus, Diacalpe, Diplazium, Dryopteris, Lastrea, Leptogramma, Peranema, Polystichum,* and *Rumohra* (Momose, 1937–1940; Mehra, 1949; Nayar, 1960; Nayar and Chandra, 1965; Nayar and Kaur, 1966) the terminal cell of the germ filament does not end in a trichome and, consequently, a lateral meristematic cell is not formed. This is apparently true whether or not the mature gametophyte possesses trichomes. Thus, the aspidioid gametophyte develops more or less symmetrically to the cordate mature form. This is in contrast to the development of the blechnoid gametophyte, which proceeds from the germ filament through more or less angular, asymmetrical stages to the mature cordate form (Nayar et al., 1966, figs. 23, 29, *Blechnum orientale;* fig. 45, *B. elongatum,* and fig. 69, *Woodwardia martinezii*).

The onocleoid gametophyte develops more or less symmetrically from the germ filament to the mature cordate form. This is the case both in those gametophytes possessing trichomes (*Onoclea, M. orientalis*) and those which are glabrous (*Onocleopsis, M. struthiopteris*).

PROTHALLIAL TRICHOMES

Atkinson and Stokey (1964:59) stated that "The matter of hairs on the fern gametophyte is at once amazing and confusing. Any pronouncement as to their phylogenetic value would be premature. . . ." So far as I am aware no monograph has appeared since the time of their publication concerned with gametophytic trichomes. The extent of our knowledge of these structures has improved only slightly, because of the more recent treatments on general gametophyte morphology.

Prothallial trichomes are of three types: (1) unicellular, glandular or aglandular (Atkinson and Stokey, 1964, figs. 27b–29): this is the type most frequently found in ferns although it may occur with other types on the same thallus. The families Aspidiaceae and Blechnaceae are characterized by this type; (2) unbranched two-celled to multicellular (op. cit., figs. 35, 43); (3) branched trichomes: not common in fern gametophytes but occurring infrequently in some genera of Grammitidaceae and Polypodiaceae, as well as in *Anemia* and *Thelypteris.*

The gametophytic trichomes of onocleoid ferns are similar to those found frequently in Aspidiaceae and Blechnaceae. Trichomes nearly identical to those of *Onoclea* have been described for several species of *Dryopteris* (Momose, 1937*a*, 1938*c*), *Acrophorus stipellatus* (Nayer and Kaur, 1966); *Athyrium esculentum* (Nayar, 1960), *A. pycnosorum* (Momose, 1939), *Ctenitis maximowicziana* (Momose, 1940), *Diacalpe aspidioides* (Nayar and Kaur, 1966), *Elaphoglossum villosum* (Stokey and Atkinson, 1957), *Lithostegia foeniculacea* (Nayer and Kaur, 1966), *Peranema aspidioides* (Nayar and Kaur, 1966), *Polystichum aculeatum* var. *taiwanense* (Momose, 1938*a*), and *Rumohra simplicior* (Kawasaki, 1956). Momose (1937*b*) has described trichomes of *Cyrtomium falcatum,* as unicellular, clavate, and glandular, and as such similar to those of *Onoclea.* However, my cultures have shown the former to be normally bicellular in structure. The only species of Blechnaceae reported with trichomes identical to those of *Onoclea* is *Blechnum brasiliense* (Nayar, 1962). Generally, the blechnaceous gametophytic trichomes are characterized by being bicellular or multicellular. This condition has been reported by Nayar (1962) for *Blechnum gibbum* and *B. orientalis*. He also reports that the larger, older trichomes on gametophytes of *Doodia* and other species of *Blechnum* have a tendency to be septate near the base. This condition is sometimes seen in older gametophytes of *Matteuccia orientalis*, although not commonly. These bicellular trichomes have been observed in my cultures as well as in those of Nayar (1961).

Multicellular and branched trichomes markedly dissimilar to those of the onocleoids have been found in the aspidioid species *Tectaria macrodonta* (Nayar and Kaur, 1964, figs. 30–32).

Thus, the type of trichome found in the onocleoid ferns occurs frequently in Aspidiaceae. The bicellular type of Blechnaceae is in contrast to this and is apparently another feature which serves to distinguish Blechnaceae from Aspidiaceae.

GAMETANGIA

Archegonia.—The general structure of the archegonium in leptosporangiate ferns is nearly uniform throughout the class (Bower, 1923). The only variable that can be utilized for comparative purposes is the number of neck cells. However, in all species studied as well as in those reported in literature, the number of neck cells in Aspidiaceae, Aspleniaceae, and Blechnaceae appears to vary only from three to seven in the three families. No unusual morphological features are present in this structure that would serve to distinguish the onocleoid ferns from the rest of the higher ferns.

Antheridia.—The structure of the mature antheridium, with a basal, ring, and cap cell, is identical in all three genera of onocleoid ferns. This is the most advanced type reported for leptosporangiate ferns and is found in Aspidiaceae, Blechnaceae, Aspleniaceae, and Davalliaceae. There are certain minor differences in antheridial structure, however, between species of Aspidiaceae and Blechnaceae which should be clarified for comparative purposes.

The major variable in the structure of the antheridium is the shape and construction of the basal cell. In onocleoid ferns and in most aspidioid ferns which have been studied (Atkinson and Stokey, 1964, fig. 49; Mehra, 1949; Nayar, 1960; Nayar and Kaur, 1964, 1966) the antheridium is globular with a more or less discoid basal cell, and the upper and lower walls are nearly parallel to one another at their edges. In many cases the wall separating the basal cell from the ring cell is centrally distended downward to a point almost touching the lower wall of the basal cell (Nayar, 1961, fig. 19, *Matteuccia orientalis*). Other aspidioid species, for example *Ctenitis maximowicziana* H. Ito (Momose, 1940), *Athyrium wardii* (Momose, 1938*d*), *Diplazium latifolium* Moore (Momose, 1938*b*), have stalked antheridia.

In *Blechnum* (Atkinson and Stokey, 1964, fig. 50a–b; Nayar, 1962, fig. 30; Stokey and Atkinson, 1952*b*, figs. 13–24; Stone, 1961, figs. 36–57) and other genera of Blechnaceae (*Stenochlaena*, Stokey and Atkinson, 1952*a*, figs. 20–22; *Doodia*, Nayar et al., 1966, fig. 99) there is a tendency for the antheridium to be elongate, with the basal cell region either multi-cellular or composed of a single, broad, distended basal cell. Some species of *Blechnum* (e.g., *B. gibbum,* Nayar et al., 1966, fig. 98) have antheridia similar in structure to those of the aspidioid species studied. This, however, appears to be rare.

The distended or multicellular basal cell region in the antheridium of Blechnaceae occurs commonly in those genera studied and appears to be a consistent character for the family. Although blechnoid-like antheridia do occur in Aspidiaceae, they are neither common nor consistent. Conversely, the typically globose antheridium of Aspidiaceae occurs infrequently in Blechnaceae. The antheridium of the onocleoid ferns is consistently aspidioid in character.

Young Sporophyte

From the little already known about sporophyte development it is apparent that it may offer fresh insight into the relations of now enigmatic fern genera. Although studies, especially of the first fronds, are still relatively few, there are indications that the earliest stages in frond development are quite similar in related genera, even in those which may have a widely divergent mature morphology. This is the situation in *Phyllitis* and *Asplenium* (Goebel, 1898).

Unfortunately, little is known about sporophyte development of ferns although modern pteridologists are beginning to emphasize this aspect in their work. Classical investigations of considerable merit have been made by Slosson (1906) and Orth (1938). More recent investigations include those on *Diellia* (Wagner, 1952), *Athyrium* (Nayar, 1960), *Lastrea* (Nayar and Chandra, 1965), *Tectaria* (Nayar and Kaur, 1964), and *Blechnum, Doodia, Stenochlaena,* and *Woodwardia* (Nayar et al., 1966).

Materials and methods.—Young fronds were taken from young sporophytes grown from spores on agar medium supplemented with an inorganic nutrient solution (see Klekowski, 1969) as well as sterilized soil. Fronds were studied and drawn from either living material or young sporophytes mounted whole and stained in Hoyer's Medium with aceto-carmine. The latter method greatly facilitates the study of epidermal pattern, stomata, early vascularization, and trichomes.

Onoclea sensibilis.—Following fertilization of the egg within the archegonium the zygote gives rise to an embryo, the structure of which is more or less similar in higher ferns. The embryo develops into a young rhizome, a root, a foot, and a first frond. In most higher ferns the root is the first organ to penetrate gametophytic tissue (Campbell, 1905), but in *Onoclea* the first frond occasionally develops more rapidly and appears before the root.

The first frond is variable in size, shape, and number of venation dichotomies, an apparent reflection of environmental conditions at the time of initiation and maturation. Generally, the larger the frond blade, the more complicated the venation pattern.

The most common condition of the first fronds is to have the veins forked only once or twice (fig. 6, *a, b*). There are therefore only two or four vein endings, and the blade is more or less obcordate. Rarely these first fronds will have three to four vein dichotomies with eight or sixteen vein endings. In all fronds examined xylem maturation occurred in an acropetal direction except for precocious maturation of cells at the point of dichotomy of a vein. In sporophytes from Japanese plants the frond blade infrequently possessed stomata on both adaxial and abaxial surfaces with guard cells 30 to 45 microns long. American plants produced first-frond blades with stomata restricted to the abaxial side, and 35 to 40 ($\mathbf{x} = 37.5$) microns long and 30 to 40 ($\mathbf{x} = 34$) microns wide.

The stipe of the first frond is about 180 microns in diameter and regularly contains only four to five tracheal strands with annular or helical tracheids. Just below the expansion of tissue at the base of the blade the vascular trace appears to consist of only one strand of tracheal cells. As the blade expands in width the number and maturation of strands increases to a point about one-third the length of the frond. From this area to the outer margin of the frond there is a gradual decrease in number of strands to two or one.

The vascular trace develops first in the rhizome and stipe and then in the root. When the rhizome is about 150 microns in length there is a continuous vascular trace from the base of the stipe through the rhizome to the root. Procambial cells proliferate in the area of the rounded rhizome apex, which has been initiated at the base of the stipe of the first frond. The first root is about 80 microns thick and appears to be enveloped by hairs similar to gametophytic rhizoids. The root cap is composed of a loosely organized mass of cells about five-cell layers and 75 microns thick. The root itself is covered by closely appressed elongate scales, giving the structure a yellowish-brown color.

The stipe bears small, glandular, uniseriate, uni- or multicellular, clavate trichomes, 45 to 55 microns long and about 20 microns wide, each with a vermiform nucleus (fig. 6, *j, k, n*). These arise superficially from the epidermis of the stipe and occur with a frequency of about six per mm length. The unicellular sporophytic trichomes are similar to the trichomes which are commonly found on gametophytes of *Onoclea*. Some larger trichomes are bicellular, with two cells of the same size; these larger trichomes can attain a length of 125 microns. Similar trichomes occur also on the lower portion of the blade of the first frond. These are similar to gametophytic trichomes when immature, although they apparently are only slightly glandular. Infrequently a larger trichome will be branched at the base; these are three-celled with a small clavate branch arising from the bottom cell. Trichomes of this type occur also around the apex of the developing rhizome.

The second frond is initiated in a spiral sequence. The stipe and blade are initiated first, followed by the root at the base of the stipe on the abaxial side. After initiation the young frond blade appears to be in an early stage in the formation of the circinate fiddlehead of the mature sporophyte frond. At this stage the young blade is only slightly reflexed backward over the stipe and apex of the rhizome. As fronds increase in size and age during ontogeny, the number of coils increases until the mature fiddlehead is formed. The second frond is always larger than the first, although here again environmental factors at the time of development play an important part. The complexity of organization of the second frond is about the same as that of the first frond. Wagner (1952) also found this to be the situation in *Diellia*. The venation pattern of the blade is open dichotomous with two to four dichotomies and four to sixteen vein endings,

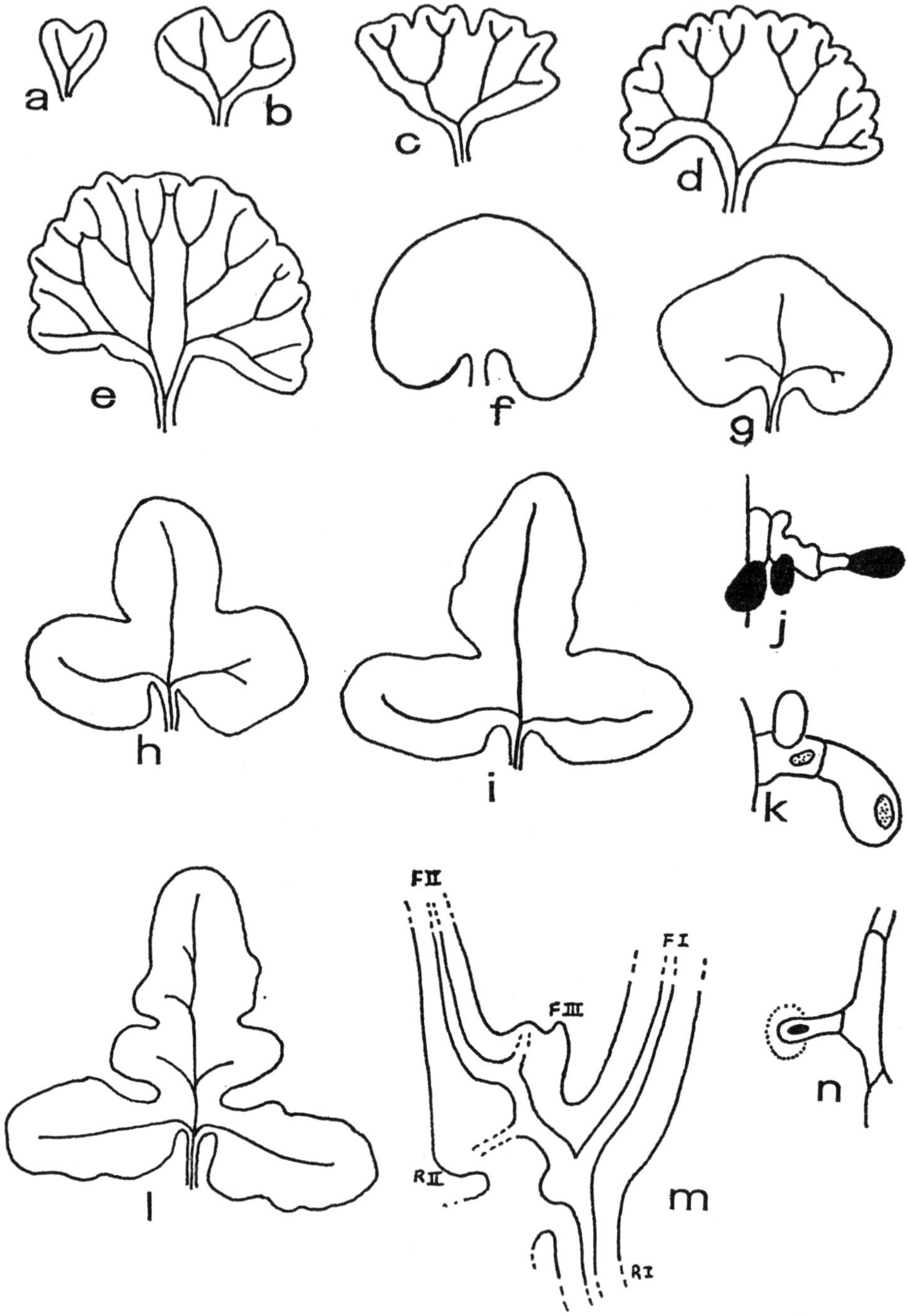

Fig. 6. Successive early fronds and sporophytic trichomes of *Onoclea sensibilis*. *a*, frond I, × 12; *b*, frond I, × 10; *c*, frond II, × 8; *d*, frond III, × 7; *e*, frond IV, × 8; *f*, *g*, frond with 7 vein dichotomies, × 2; *h*, frond with 8–10 vein dichotomies, × 1.7; *i*, frond with 12–13 vein dichotomies, × 1.3; *j*, trichome on young stipe, × 167; *k*, trichome on young stipe, × 300; *l*, frond with 12–13 vein dichotomies, × 1.1; *m*, young rhizome with two fronds and a frond primordium, × 35; *n*, simple stipe trichome, × 200.

although this varies with size of frond, smaller fronds having fewer dichotomies and hence fewer vein endings (fig. 6, *c*). The apex of the rhizome at this stage continues to be covered by numerous uniseriate and branched two- to six-celled hairs with a greatly enlarged terminal cell. The blade of the second frond bears a few scattered, uniseriate, one- to three-celled hairs.

The young lamina of the third frond, as in the second frond, is more or less circinate with the blade bent back over the apex.

The first meristematic cells of the third frond are initiated when the second frond is about 1.5 mm long. The young lamina grows rapidly and quickly surpasses the rhizome apex in length. The mature blade of the third frond has open dichotomous venation with three to four dichotomies and eight to sixteen vein endings (fig. 6, *d*). Large trichomes with two to four cells are frequent surrounding the rhizome apex. These larger trichomes posses a large expanded terminal cell about 38 to 70 microns long and a smaller basal cell 20 to 23 microns long. Many of these larger trichomes are branched at the base.

The fourth frond is the earliest frond to show an occasional vein anastomosis, invariably between veins lying adjacent to the center of the blade near the outer margin. The blade is about 4 mm in breadth with four to five vein dichotomies. The veins contributing to the anastomosis are of the two different dichotomous systems traced back to the original dichotomy of the midvein of the stipe. Thus, the first areole is always central, running almost the entire length of the blade between the two dichotomous systems. Anatomoses do not occur in American material until five vein dichotomies are present in the blades. Here again, the first areole is usually centrally located.

Trichomes on the stipe of the fourth frond are more complex in structure than those found on earlier fronds. These unicellular trichomes present are identical with gametophytic trichomes.

The fifth and later fronds show a gradual transition toward monopodial branching in the venation system with a change in pattern from two equal dichotomous systems to a pattern with a definite midvein (fig. 6, *f–i*). The first midvein appears in fronds 20 to 25 mm wide with seven to eight vein dichotomies from the base. A vein from the second dichotomy assumes the function of the midvein, extending into the center lobe of the blade. Increase in frond size appears to be most rapid during this phase of ontogeny. The blades become larger and more complex with numerous anastomoses and with the gradual appearance of pinnae lobes near the base of the center pinna. Fronds with six vein dichotomies show about seven anastomoses. Those with eight to ten dichotomies show consistent anastomoses and long costal areoles near their base.

The anastomoses in young fronds of *Onoclea* may be classified into several different types according to the system presented by Foster (1966, 1968). These include the following:

Type I. Transient union between adjacent branches of two-vein dichotomies. There is fusion of the two veins for a short distance and then divergence from one another.

Type II. Permanent union of the inner branches of two adjacent vein dichotomies into a single excurrent vein.

Type III. Interconnection of two closely adjacent vein dichotomies by a very short, obliquely oriented commissural vein. The two branches of the vein dichotomies remain distinct.

Type IV. Transient union between an unbranched arcuate vein and the branch of a contiguous vein dichotomy (Foster, 1968: fig. 6). This is similar to Type I except for the veins involved.

Table 4 summarizes the types found and their frequencies in young fronds of *Onoclea*.

Data on frequencies of various types of anastomoses in *Onoclea* indicate that in these young frond blades, which show the transition from open to reticulate venation, the structural union of two veins is accounted for in nearly all cases

TABLE 4

TYPE AND FREQUENCY OF VEIN ANASTOMOSES IN YOUNG FROND BLADES OF ONOCLEA SENSIBILIS

Total no. anastomoses/frond	Number and type of anastomoses				Other
	I	II	III	IV	
5	1	2	1	1	0
7	1	3	1	0	2
19	2	13	4	0	0
26	6	12	8	0	0

by type I, II, and III. Of these type II is found most frequently in all fronds of all ages. However, as fronds become larger the frequency of type II increases considerably.

Foster (1968) has found that type I and II anastomoses occur most frequently in the venation of *Anemia herzogii* Rosenstock and he thinks that these types "seem to represent extremely ancient trends in the evolution of areolate venation" (Foster, 1968: 63). He also notes that these types of anastomoses occur in the fossil genera *Neuropteris, Ctenis, Gangamopteris,* and *Psygmophyllum flabellatum,* as well as in the living Angiospermous herb *Circaeaster agrestis* Maxim.

In earlier fronds of *Onoclea* anastomoses occur exclusively in the outer tier of veins. However, as fronds increase in size with more numerous veins, anastomoses occur with greater frequency between veins of the second tier of dichotomies from the margin of the frond. Fronds with eight to ten dichotomies show consistently the long costal areoles so typical of the mature frond.

As the fronds gradually increase in size there is a correlated appearance of lobes on either side of the reniform or elongate pinna near the base. This usually is initiated in the fifth frond. Stomatal size in fronds of this age varies from 32 to 42.5 ($\bar{x} = 39.0$) microns in length and 22.5 to 30 ($\bar{x} = 26.0$) microns in width. All trichomes on the stipe and lower portion of the main veins of the lamina are of the multicellular sporophytic type, and no unicellular trichomes of the gametophytic type were observed.

In the sixth frond with 12 to 13 vein dichotomies the establishment of long costal areoles is well fixed. The midvein of the terminal central pinna develops much sooner than the midvein of the lateral pinnae. The costal areoles extend for two-thirds to three-fourths the length of the central pinna. These areoles are just beginning to appear in lateral pinnae.

In the seventh and later fronds the pinnate nature of the frond is fully achieved by the gradual lobing of lateral areas of the central pinna, followed by establishment of a midvein with costal areoles in each lobe. Stomatal size in these fronds varies from 40 to 42.5 ($\mathbf{x} = 41.25$) microns long and 27.5 to 30.0 ($\mathbf{x} = 29.6$) microns wide.

The primary change in frond structure from the seventh frond to maturation is an increase in size and additional lobing of the terminal pinna.

Onocleopsis hintonii.—The structure and development of the embryo of *Onocleopsis* is similar to that of *Onoclea* and most higher ferns. Gametophytes of this species continue to grow and produce new archegonia after fertilization and the development of the embryo. Polyembryony has not been observed. When the embryo is about 400 microns long it consists of foot, shoot apex, frond primordium, and root apex. At this stage procambial tissue has developed throughout the main axis from the frond to the shoot apex and root apex. No procambial tissue extends into the foot. The first root grows basally simultaneously with the growth of the first frond toward the notch of the gametophyte. When the embryo is 800 microns long, xylem elements with helical thickenings are present along the shoot-root axis. At this stage new archegonia are still being produced in the notch region of the gametophyte and new antheridia occur in its wings. The apex of the first frond is covered with unusual unicellular protuberances or trichomes 35 to 65 microns long.

The stipe of the first frond bears two types of trichomes. These are (1) trichomes resembling gametophytic trichomes of *Onoclea* and similar to simple sporophytic trichomes on the first frond of *Onoclea* (fig. 7, *j*) and (2) compound trichomes identical in structure to those of *Onoclea*. The unicellular trichomes are clavate, 75 to 90 microns long and scattered infrequently along the stipe and frond blade. Occasionally, they may be bicellular with a shorter, broader terminal cell and a longer, more or less collapsed basal cell. These bicellular trichomes are about 140 microns long and extend along the vascular trace of the blade only to the point of the first vein dichotomy. They also occur occasionally on the margin of the blade. The unicellular trichomes apparently represent developmental stages of the bicellular trichomes, because all intermediate stages between the two types occur. As the unicellular trichomes elongate a transverse wall is laid down near their distal end. The basal cell then appears to collapse, due to loss of turgor and cytoplasmic contents, which have concentrated in the terminal cell. The terminal cell is golden-brown in color.

The bicellular trichomes of *Onocleopsis* rarely branch near the base, a condition in contrast to that found in *Onoclea*. As the trichomes are produced early in ontogeny and are mature when the frond primordium is no more than 250 microns long they probably serve in a protective capacity.

When the first root is 1.5 mm long it has a root cap about 37 microns thick and is covered by brownish, unicellular rhizoids similar to those found on roots of *Onoclea* and *Matteuccia* spp.

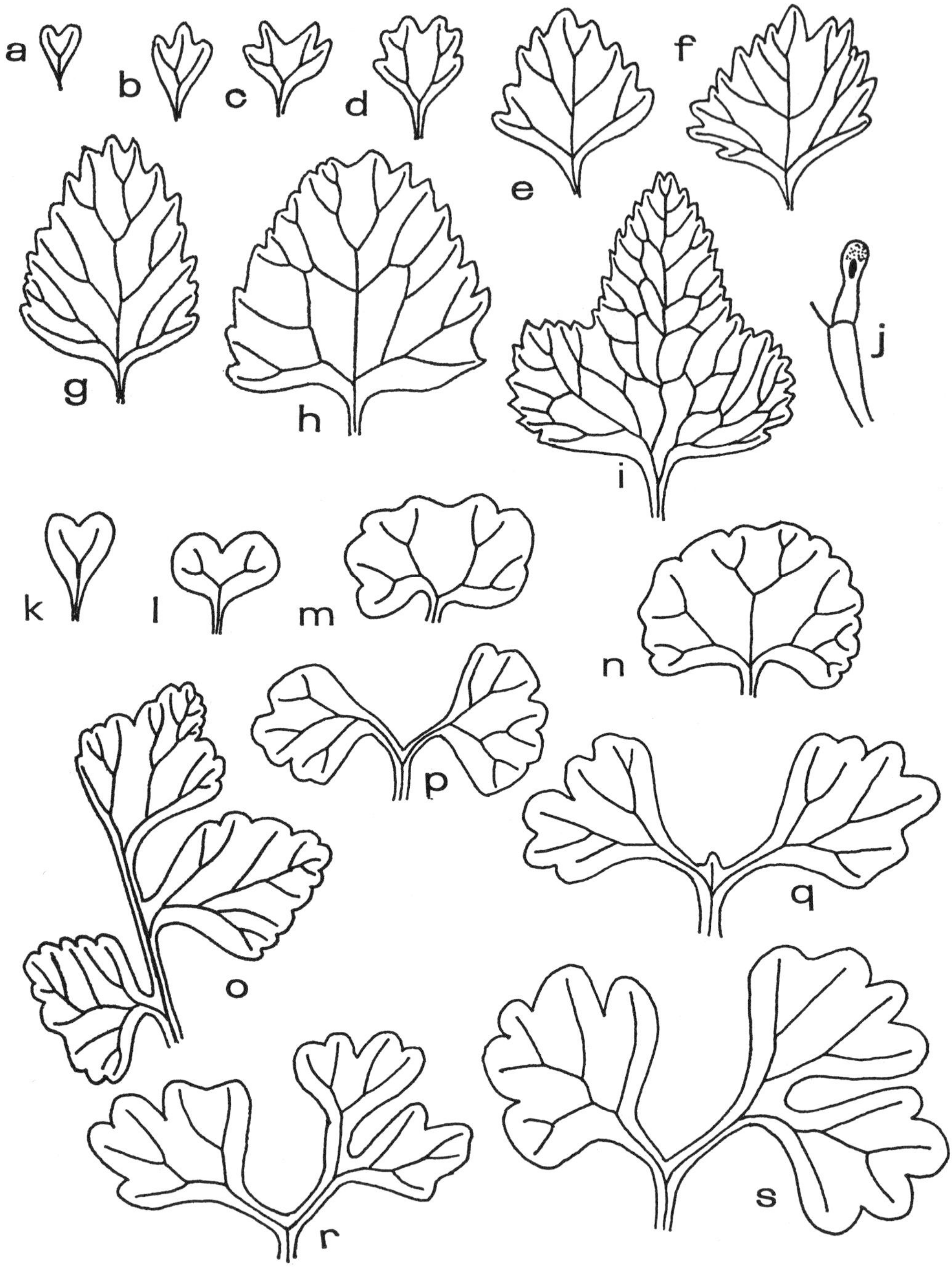

Fig. 7. *a–j,* successive early fronds of *Onocleopsis hintonii; k–o,* successive early fronds of *Matteuccia orientalis; p–s,* early fronds of *Matteuccia struthiopteris. a,* frond I, × 1.6; *b, c, d,* frond II, × 1.7; *e,* frond III, × 1.7; *f,* frond IV, × 1.7; *g,* frond V, × 1.7; *h,* frond VI, × 1.7; *i,* frond VII, × 1.7; *j,* stipe trichome, × 160; *k,* frond I, × 13; *l,* frond II, × 6; *m,* frond III, × 6; *n,* frond IV, × 6; *o,* frond II, × 2.5; *p,* frond I, × 2; *q,* frond II, × 2; *r,* frond III, × 3; *s,* frond IV, × 4.

Immediately after initiation the blade of the first frond bends back over the stipe and shoot apex at an angle of about 90 degrees. After development the blade is characterized by open-dichotomous venation with only one (rarely two) dichotomy at the base of the blade (fig. 7, *a*). The blade attains a maximum length of about 4 mm with a stipe 8 to 10 mm long.

Stomata are found infrequently on the abaxial side of the young blade. Guard cells are 35 to 53 ($\mathbf{x} = 45$) microns long and 28 to 38 ($\mathbf{x} = 34$) wide.

At this stage in the development of the first frond, the second frond has been initiated and is about 1.5 mm long. Procambial tissue is well differentiated and the blade exhibits a simplified circinate structure similar to that of the young frond of *Onoclea*. Nuclei of the procambial cells are elongate, 17 to 35 microns long and 3 to 6.5 microns wide. Nuclei of nonprocambial cells are spherical, 8 to 13 microns in diameter. Differentiation of xylem elements is acropetal in sequence. The root primordium of the second frond is well differentiated at this stage. Primordial cells for the third frond and root are present. The apices of both second and third fronds are covered with bicellular trichomes similar to those found on the first frond.

The mature blade of the second frond is about 9 mm long and cuneate. The venation pattern is open-dichotomous with two to three vein dichotomies and three to six vein endings (fig. 7, *b*, *c*, *d*). Both unicellular and bicellular trichomes are found on the stipe and most of the blade. Trichomes of the stipe commonly arise from either the basal or apical end of the epidermal cells. On the blade, trichomes are restricted to the epidermis immediately above the veins. These bicellular trichomes are 88 to 200 microns long and 30 to 45 microns wide. In most cases the basal cell is slightly longer than the terminal cell.

Stomata are present on the abaxial side of the frond. These are 32 to 45 ($\mathbf{x} = 37$) microns long and 30 to 40 ($\mathbf{x} = 34$) microns wide.

During the ontogeny of the second frond there is considerable upward elongation of the rhizome into an erect attitude. Fronds are then initiated spirally, as is typical of ferns with erect rhizomes.

The primordium of the third frond appears shortly after the initial elongation of the second frond, when the latter is about 1 mm long. The apical region of the rhizome as well as the veins and stipe of the second frond are covered with bicellular trichomes.

The venation pattern of the third frond is open-dichotomous with four to six vein dichotomies and nine to seventeen vein endings (fig. 7, *e*). In some instances the second dichotomy from the base appears to establish a monopodial venation pattern. The stipe and veins of the immature third frond are covered with numerous large bicellular trichomes 95 to 115 microns long. The mature blade of the third frond is mostly devoid of trichomes.

The venation of the fourth frond is characterized by a definite midvein with lateral, dichotomously branching veins (fig. 7, *f*). There are about fifteen vein endings in the blade. The stipe bears an occasional bicellular trichome about 150 microns long. Stomata on the blade are ovate in shape, 45 to 60 ($\mathbf{x} = 55$) microns long and 30 to 40 ($\mathbf{x} = 35$) microns wide.

The structure of the fifth frond is similar to that of the fourth, although it is slightly larger (fig. 7, *g*).

The sixth frond is characterized by the appearance of costal areoles, usually two, on either side of the midrib near the base of the blade (fig. 7, *h*). The areoles are produced by anastomosis of branches of two different vein dichotomies from lateral veins arising from the midvein at different distances from the base of the blade. Occasionally, larger sixth fronds will produce four costal areoles, one above the other on either side of the midvein. The sixth frond is mostly glabrous except for an occasional bicellular trichome on the stipe.

The seventh frond is characterized by the appearance of lateral lobes and frequent vein anastomoses (fig. 7, *i*). Stomata are 40 to 45 microns long and 35 to 40 microns wide.

The eighth frond shows progressive development of lateral lobing. The ninth and succeeding fronds exhibit a further development of lobing of the central pinna-lobe, with progressive anastomoses of veins throughout the blade. This process results in the mature pinnately compound frond.

Matteuccia struthiopteris.—The first frond of *M. struthiopteris* is invariably pinnate, normally with two more or less cuneate pinnae resulting from a dichotomy of the stipe near its apex but below the point of blade-tissue production (fig. 7, *p*). This frond structure is basically similar to that found in the first fronds of *Onoclea, Onocleopsis,* and *M. orientalis* except for the lack of blade tissue at the point of the lower dichotomy of the stipe. Rarely, a third pinna will be formed on the first frond, resulting from the dichotomous branching of one of the two dichotomous branches of the stipe. In this case, the frond then appears identical to the third frond. The venation is open-dichotomous. Each pinna of the first frond bears two to three vein dichotomies with six to eight vein endings.

The mature blade of the first frond is glabrous. Bicellular or multicellular trichomes occur occasionally on the stipe. The terminal cell of these trichomes is similar to that of the bicellular trichomes in *Onoclea* and *Onocleopsis*. However, trichomes in *M. struthiopteris* may be uni-, bi-, or multicellular. In addition, scales may be found near the base of the stipe. These structures are deltoid to lanceolate with a terminal cell similar in all respects to that of the trichomes. The trichomes and scales may represent two aspects of the same ontogenetic series, since many intermediates between the two extremes can be found. The terminal cell is golden-brown in color and varies from 80 to 150 microns long; it is clavate and slightly curved. The basal cell of the trichome is of variable length and structure. The scales at the base of the stipe are about 0.5 mm long and 0.1 mm wide. Scales and large, uni- or multicellular golden-brown trichomes are frequently found around the apex of the rhizome and of the young frond primordia.

Stomata on the first frond are round to ovate and 30 to 33 microns wide and 35 to 38 microns long.

The second frond is pinnate. Its pinnae are cuneate-ovate with three to four vein dichotomies and six to eight vein endings (fig. 7, *q*). Stomata are similar in size and shape to those found on the first frond. The blade of the mature second frond is glabrous.

The stipe of the third frond bears occasional bicellular clavate trichomes similar to those of the first frond. The blade is normally bipinnate and completely glabrous (fig. 7, *r*).

Frond four is similar to frond three except that it is slightly larger. At this stage there is no sign of formation of a definite midvein (fig. 7, *s*).

No observations were made on later fronds in this series.

Matteuccia orientalis.—The first frond of *M. orientalis* is usually cuneate to ovate (fig. 7, *k*). Its venation pattern is open-dichotomous and varies from one vein dichotomy with two vein endings in the simplest fronds to four vein dichotomies with about fifteen vein endings in larger and more complex fronds. Rarely, first fronds will appear to have a midrib caused by the apparent trichotomy of the main vein at the base of the blade. This condition was also reported by Nayar (1961). The rhizome is erect at this stage of development. Structure and development of rhizome and root is more or less similar to that of *Onoclea* and *Onocleopsis* and will not be discussed here.

Stomata on the blade of the first frond are ovate to round, 40 to 48 microns long and 30 to 40 microns wide.

The region surrounding the rhizome apex, the young stipe, and the lower veins of the blade are covered with uni- and bicellular trichomes similar to those found in *Onoclea* and *Onocleopsis.* These are clavate and 70 to 150 microns long, with a large, expanded terminal cell and a shorter basal cell. The basal cell varies in length from 15 to 30 microns. The terminal cell is golden-brown and 55 to 120 microns long.

The structure of the second frond is correlated directly with the complexity of the first frond. If the first frond is of the simplest type then the second will usually possess two vein dichotomies and four vein endings (fig. 7, *l*). If the first frond is characterized by a midrib then the second will frequently be pinnate (fig. 7, *o*). The compound pinnate condition is established early in the ontogeny of the frond. This is reminiscent of the early establishment of pinnation in *M. struthiopteris.* Fronds subsequent to the second show proliferation of the basal area of the terminal pinna. Stomata on the blade of the more complex second frond are ovate to round, 45 to 50 microns long and 30 to 40 microns wide. Trichomes of the second frond are rare on the veins of the lower portion of the blade; the stipe bears an occasional unicellular trichome. Bicellular trichomes are frequent on the young, unrolled blade and around the shoot apex.

No observations were made on later fronds in this series.

Summary of the onocleoid ferns.—Development of the frond of the young sporophyte in all three genera is characterized by the early appearance of open-dichotomous venation or branching in simple, more or less cuneate fronds. In later fronds, monopodial venation is established by formation of a midvein by one of the veins of a lower vein dichotomy. The establishment of monopody in an occasional first frond of *Matteuccia orientalis* is unusual in the group. *Onoclea* and *Onocleopsis,* with reticulately veined mature fronds, are further characterized by anastomoses of veins and the appearance of long costal areoles in early fronds, later followed by areole formation throughout the blade.

The three genera are similar in regard to structure of trichomes; these are essentially similar in development and morphology in *Onoclea, Onocleopsis,* and *Matteuccia orientalis.* The ultimate development of scales is a further elaboration found in *M. struthiopteris.*

Mature Sporophyte

ROOTS

Roots of onocleoid ferns are generally coarse, black, and wiry, and are frequently produced at the base of stipes, arising therefore from between the persistent stipe bases on the rhizome. Roots of *Onoclea* arise near the base of the stipe on the dorsiventral rhizome and are 0.1 to 1.0 mm in diameter. Root hairs appear throughout much of their length and are fine, white, and up to 1 mm long. Roots of *Matteuccia* and *Onocleopsis* are brownish when young, because of a thick covering of brown to dark brown, linear, unicellular hairs about 350 to 700 microns long and 24 microns wide. Older roots have lost much of their covering of hairs, are black in color, and are 0.6 to 2.0 mm in diameter.

The internal root anatomy of *Onoclea Onocleopsis,* and *Matteuccia* is very similar and root anatomy in ferns is fairly constant. The epidermis is one cell thick and composed of cells with thick lignified secondary walls. These cells are 15 to 20 microns in diameter. The cortex in *Onoclea* is four to six cells thick, and its outer cells have thick secondary walls, the thickness of the secondary walls decreasing toward the stele. The cortex of *Matteuccia* is ten to twelve cells thick and with similar secondary-wall thickenings. The endodermis in all onocleoid ferns consists of small, thin-walled cells without suberized walls. The stele varies in thickness from about 95 to 180 microns, depending upon the overall diameter of the root. Xylem is diarch, with two narrow zones of protoxylem, one at either outer end of the metaxylem. The remaining tissue of the stele consists of a rather prominent pericycle, phloem, and parenchyma cells.

Comparisons.—Root anatomy generally has been ignored by pteridologists in monographic studies. This is no doubt because of the belief that root anatomy offers little in the way of comparative characteristics and the very small amount of information in the literature. These thoughts were expressed by both Wagner (1952) and Ogura (1938). The major variable appears to be degree of sclerification of the cortex. Wagner (1952) found that in mature roots of *Diellia* most cortical cells are heavily sclerified. Rao and Khare (1964) reported a heavily sclerified inner cortical region in *Tectaria amplifolia* (V.A.V.R.) Chr. My investigations of *Blechnum falciforme* and *Woodwardia areolata* indicate a basic structural similarity between these genera and onocleoid ferns except for a greater amount of sclerification of the inner cortical cells. This condition is apparently typical for many advanced ferns (Schoute, 1938).

RHIZOME

The rhizome in onocleoid ferns varies from being erect and trunklike in *Matteuccia* and *Onocleopsis* to prostrate and dorsiventral in *Onoclea.* The rhizome in *Onoclea sensibilis* is widely creeping, flattened, and more or less black in color; it is about 2 to 5 mm in lateral width and 9 to 12 mm in vertical width. Older fronds are spaced about 1.5 cm apart, whereas newer fronds produced near the apex are crowded close together. The rhizome is typically dorsiventral with fronds arising from the upper portion and roots from the lower. There is apparently no relation between origin of root traces and position of fronds, a situation found commonly in other dorsiventral prostrate rhizomes. Stipe bases are more or less persistent.

The rhizome of *Onoclea* is mostly glabrous except for a few broadly ovate brown scales covering the apex and scattered linear-lanceolate, black scales about 1 cm long and 8 mm wide, with bases profusely fimbriate with reddish-brown capillary outgrowths (fig. 9, *j*).

Internally the vascular system is dictyostelic, a condition uniform throughout advanced ferns (fig. 8, *a–d*). The epidermis of the rhizome is scarcely distinguishable from the outer cortex since both are highly lignified with thick secondary cell walls. The outer three to four cell layers of the cortex are uniform in size and structure. Inward the parenchyma cells gradually increase in size and are more or less uniform throughout the ground tissue except in the immediate vicinity of the meristeles, where they become smaller. In my sections the ground tissue cells were filled with abundant small, round-ovoid starch grains about 2.5 to 8.5 microns in diameter. Parenchyma cells nearest the meristeles were more densely protoplasmic with smaller starch grains.

Vascular bundles, or meristeles, are amphicribal concentric-oval strands, about 150 to 190 microns in diameter. The endodermis surrounding the bundle is composed of thin-walled, more or less rectangular cells smaller than either the inner cortical cells or the cells immediately within the meristele. The bundle itself consists of a lens- or oval-shaped band of xylem about 250 microns wide surrounded by phloem. The outer region of the bundle has numerous parenchyma cells which appear to function in storage. Leaf gaps are apparently quite broad.

In contrast to *Onoclea* the rhizome of *Matteuccia* is erect and densely covered with persistent stipe bases from which emerge numerous black roots. Erect rhizomes of *M. struthiopteris* frequently produce long black lateral prostrate rhizomes 2 to 3 mm in diameter with numerous black roots arising from their bases. These lateral rhizomes run out in several directions from the parent plant one to several inches under the ground and at varying intervals produce new plantlets. They frequently persist after production of a new plant. It seems possible, therefore, that some populations consist of a single clone.

The rhizome of *M. struthiopteris* is more or less dauciform about 10 cm long and 1 to 3 cm in diameter (without stipe bases). The vascular system is dictyostelic and the fronds arise spirally. The outer cortical cells are sclerified with dark reddish-brown walls. This tissue grades abruptly into the normal parenchyma cells of the remaining cortex. The meristeles of *Matteuccia* are similar in structure to those of *Onoclea.*

The rhizome of *Matteuccia orientalis* is broader than that of *M. struthiopteris.* In addition it lacks the profuse stolons of *M. struthiopteris,* although lateral branches do appear near the base of the rhizome, apparently from adventitious buds. Nayar and Kazmi (1963) report that the cells of the ground tissue are more or less thick-walled, the wall thickening being more prominent near the center of the rhizome than near the epidermis. They also report a stelar cylinder highly dissected by broad, obovate leaf gaps; at each gap a pair of broad strands originates and traverses the cortex into each stipe. Hayata (1928) figured five root strands arising along the basal region of the leaf gap in *M. orientalis.*

The rhizome of *Matteuccia intermedia* has been investigated previously by Bower (1914) who cultivated this species at the Glasgow Botanic Garden. The rhizome is an open dictyostele with a very bulky pith and with leaf traces originating as a pair of strands separate from the beginning. From Bower's description and figure there apparently is little variation between the structure of this and other *Matteuccia* species.

The rhizome of *M. struthiopteris* has hollow spaces immediately internal to departure of the leaf bases. This ring of air space immediately within the ring of stelar bundles results from the shape of the leaf base and its apparently long close association with the rhizome before divergence.

The rhizome of *Onocleopsis hintonii* is erect and trunklike, up to 1 meter tall and 5 to 20 cm in diameter including the persistent leaf bases. Short lateral rhizomes are occasionally produced near the base; these are about 2 cm in diameter and terminate in new plantlets. Commonly, rhizomes which have been dislodged from their normal upright position will show an upward curvature of the apex. Wagner (1952) also reported this situation in *Diellia.* Fronds are produced spirally and appear in a crownlike fashion, similar to the appearance of fronds in *M. struthiopteris* and *M. orientalis.* Leaf-gap regions, as in *Onoclea* and *Matteuccia,* produce two strap-shaped strands from different meristeles which traverse the cortex into the base of the stipe.

Comparisons with other genera.—The occurrence of both erect and prostrate rhizomes in a related group of plants is not unusual. For example, both types occur within the genera *Blechnum* and *Woodwardia* (Copeland, 1947). This is also true in *Asplenium* (Wagner, 1954) and in the family Aspidiaceae. There is thus evidence that this feature has arisen more than once in very closely related groups of ferns. Hence, the presence of both erect and prostrate rhizomes in onocleoid ferns has little phylogenetic significance.

The variability in the closely related families Aspidiaceae and Blechnaceae makes it difficult to find constant comparative characteristics. Ogura (1938), however, separated the genera associated commonly in Blechnaceae and the onocleoid genera into two separate tribes of Polypodiaceae *sensu lato,* citing the type of rhizome stele as one distinguishing character. In Woodsieae, with a typical or perforated dictyostelic rhizome, he placed *Diacalpe, Peranema, Woodsia, Hypoderris, Cystopteris, Acrophorus, Matteuccia,* and *Onoclea.* In his tribe Asplenieae, with a *Cyathea-* or *Stenochlaena*-type dictyostele he placed the genera *Athyrium, Diplazium, Phyllitis, Boniella, Camptosorus, Asplenium, Blechnum, Brainea, Stenochlaena, Woodwardia,* and *Doodia.* He also found correlations with the occurrence of mechanical tissue. Here he figured three types: (1) mechanical tissue present only as a hypodermal ring (including *Onoclea, Matteuccia, Diplazium,* and *Sadleria cyatheoides*); (2) mechanical tissue forming a hypodermal ring and a small mass near the meristeles (including *Woodwardia radicans,*

Dryopteris sophoroides, and *Diplazium esculentum*); and (3) mechanical tissue comprising a great mass in cortex and medulla (including *Blechnum nipponicum).*

My investigation of the rhizome of *Woodwardia areolata* revealed a structure similar to that of *Onoclea,* with a less sclerified outer cortical region and an inner ground tissue of parenchyma.

It is apparent from this discussion that the rhizomes of Aspidiaceae and Blechnaceae differ only in minor details. Those differences which are readily apparent, however, may be due to environmental modifications of stress and support, and thus probably have little value in determining phylogenetic relationships.

STERILE FROND

Structure of the stipe in both sterile and fertile fronds of *Onoclea, Onocleopsis,* and *Matteuccia* is similar and differs in only minor details in each genus. The stipe of the sterile frond is stramineous with a dark brown to black base. The abaxial side is rounded to obtusely angled, the adaxial side flattened to grooved. At the base the stipe is densely covered with scales, these diminishing in abundance toward the apex of the frond. The stipe of *Onoclea* is 3 to 4 mm wide at the base, those of *Onocleopsis* and *Matteuccia* are broader, up to 1 cm in *Matteuccia* and 2 cm in *Onocleopsis.* Stipe and fronds of *Onoclea* arise from the upper side of the dorsiventral prostrate rhizome. In *Onocleopsis* and *Matteuccia* they arise spirally from the erect rhizome.

Internal anatomy.—The stipe of onocleoid ferns contains two large strap-shaped vascular bundles, each arising from separate meristeles in the rhizome and traversing the stipe in a parallel manner; unbranched and free throughout its length, fusing at the base only further up in the rachis. (fig. 8, *e–p*). The structure of the base is relatively simple, with a thin epidermis, a uniform cortex of moderate-sized parenchyma cells, and the vascular bundles. The outer six to ten cell layers of the cortex are moderately sclerified in *Onoclea.* In *M. struthiopteris* the sclerified region can be divided into an outer region of five to six cell layers with thick, highly lignified secondary walls, and an inner region six to seven cells thick with unlignified secondary walls. The cortex of *Onocleopsis* contains an outer sclerified region eight to twelve cell layers thick, abruptly grading into the thin-walled parenchyma of the inner cortex. External to the outer sclerified cortical region of the cortex is a layer about two cells in thickness of thin-walled parenchyma, and the epidermis. In *M. orientalis* the epidermis bears a thick cuticle. It is difficult to distinguish the epidermis from the outer cortical region. The outer cortex is seven to ten cell layers thick, becoming thinner at the corners of the stipe, with cells only about one-third the size of the inner cortical cells with walls only slightly thickened.

The vascular strands of the stipe in all three genera are narrowly to broadly oblong, surrounded by a thin-walled endodermis with short rectangular cells. As seen in transverse section the pattern of xylem in each strand is linear with hooks at either end pointing toward the axis of the stipe. In *Onocleopsis* and *Matteuccia* the xylem strand is more or less uniform in thickness throughout its length.

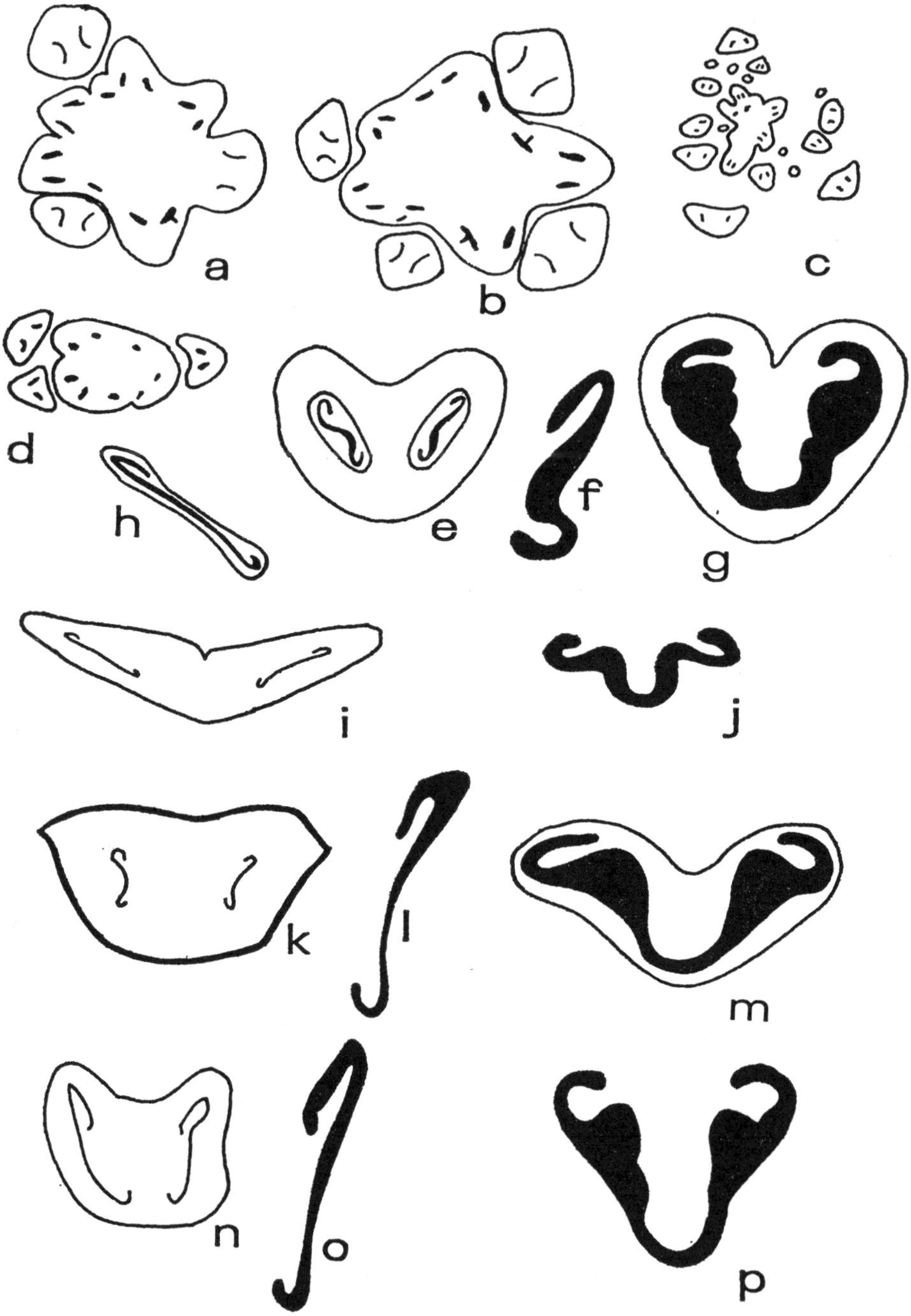

Fig. 8. Transections of rhizomes and stipes. *a, b, Onocleopsis hintonii,* young rhizome, × 1.3; *c, d, Matteuccia struthiopteris,* young rhizome, × 1.3; *e–g, Onoclea sensibilis: e,* lower stipe, × 15; *f,* xylem strand of lower stipe, × 40; *g,* vascular bundle of rachis, × 70; *h–j, Matteuccia struthiopteris: h,* vascular bundle of lower stipe, × 1; *i,* lower stipe, × 4; *j,* xylem pattern in rachis, × 25; *k–m, Matteuccia orientalis: k,* lower stipe, × 4; *l,* xylem strand of lower stipe, × 7; *m,* vascular bundle of rachis, × 100; *n–p, Onocleopsis hintonii: n,* lower stipe, × 1.6; *o,* xylem strand in lower stipe, × 3; *p,* xylem pattern in rachis, × 30.

In *Onoclea,* the central portion of the strand is considerably enlarged, giving the overall strand somewhat of an E shape. The strand is not surrounded by either a sclerified or a cuticular region in any of the genera.

Surrounding the xylem strand is an area of phloem and parenchyma cells, the latter apparently containing numerous starch grains.

Comparisons with other genera.—Major variables in the internal structure of the stipe include number of vascular strands and presence or absence of a sclerified or cutinized sheath surrounding the strand. In order to understand the evolution of the double-stranded nature of the stipe of the onocleoid ferns and other genera it is necessary to examine this structure in dennstaedtioid ferns.

Wagner (1969) and Holttum (1949) largely agree that the dennstaedtioid ferns, including *Dennstaedtia, Cyathea, Dicksonia,* and related genera, are the more primitive elements of the large complex including the families Dennstaedtiaceae, Cyatheaceae, Lindsaeaceae, Hymenophyllaceae, Aspidiaceae, Aspleniaceae, Blechnaceae, and Davalliaceae. Indeed, Holttum (1949) included all these families in his broadly conceived Dennstaedtiaceae.

The basic anatomy of the stipe in this group is characterized as omega-shaped, as exhibited by species of *Dennstaedtia, Paesia, Hypolepis, Pteris, Loxsomopsis, Culcita,* and *Saccoloma* (see Keating, 1968, for a complete discussion of vascular patterns in this group). Keating traces a progressive specialization of the vascular tissue from a primitively unbroken, complete, omega-shaped strand to more advanced conditions entailing dissection of the strand into separate parts with loss of portions of the strand. Genera intermediate in strand construction between the omega-type and the more advanced types exhibited by Blechnaceae and Aspidiaceae include *Blotiella,* with the loss of much of the abaxial portion of the trace, and *Lonchitis* with two traces oriented in much the same way as in thelypteroid and onocleoid ferns. There is a high degree of simplification in stipe structure in *Asplenium, Blechnum, Cheilanthes, Polypodium, Thelypteris, Trichomanes, Onoclea, Matteuccia,* and most other genera of more advanced families.

There apparently are two primary lines of simplification. The first is toward complete elimination of most of the original omega-type strand, resulting in a double-stranded condition. The other leads to a more complicated vascular pattern of several strands in a U shape within the stipe. Genera which have attained the double-stranded condition include *Onoclea, Onocleopsis, Matteuccia, Asplenium, Cystopteris, Athyrium, Diplazium, Woodsia, Hyalotricha, Lonchitis, Cyclosorus, Lastrea, Goniopteris, Meniscium, Glaphropteris, Steiropteris, Thelypteris,* and *Woodwardia* (*W. areolata*). Genera which have three or more discrete bundles in the stipe include *Arachniodes, Bolbitis, Ctenitis, Cyclopeltis, Didymochlaena, Dryopteris, Elaphoglossum, Lastreopsis, Plecosorus, Polybotrya, Polystichum, Tectaria, Lomariopsis, Blechnum, Nephrolepis, Oleandra, Brainea, Woodwardia, Sadleria, Doodia, Stenochlaena, Peranema, Microsorium, Acrophorus,* and *Diacalpe.* Data on most of the above genera were compiled by Alan R. Smith during 1967 in Costa Rica.

Another useful comparative character derived from stipe anatomy is the presence or absence of sclerenchymatous thickening surrounding the vascular bundles. Sclerenchymatous sheaths may be completely absent, partially present, or completely sheathing the bundle. Genera lacking a sclerenchymatous sheath include *Onoclea, Onocleopsis, Matteuccia, Athyrium, Diplazium* (one species), *Cyclosorus, Lastrea* (several species), *Goniopteris, Cystopteris, Woodsia, Dryopteris* (*D. bangii, D. dentata*). Genera with a partially developed or incomplete sheath include *Polybotrya* (*P. cervina*), *Polystichum* sp., *Diplazium* spp., *Lastrea* (*L. rudis, L. mercurii*), *Meniscium* (*M. serrata*), *Thelypteris* sp., *Blechnum* (several species). Genera with a complete sclerenchymatous sheath include *Arachniodes, Bolbitis, Ctenitis, Cyclopeltis, Didymochlaena, Dryopteris, Elaphoglossum, Plecosorus, Polybotrya (P. osmundacea), Polystichum (P. platyphyllum), Tectaria, Diplazium* spp., *Thelypteris (T. torresiana), Nephrolepis, Oleandra, Blechnum (B. occidentale, B. gibbum), Woodwardia.*

In summary, those genera similar to the onocleoid ferns in stipe structure include *Athyrium, Cyclosorus, Cystopteris, Diplazium* (in part), *Goniopteris, Lastrea* (in part), and *Woodsia.*

The striking fact about the above data is that those genera which have traditionally been recognized as allied tend to have the same type of construction of the stipe. This includes those species recognized in Blechnaceae and Davalliaceae, as well as such dryopteroid ferns as *Dryopteris, Polystichum, Ctenitis, Plecosorus, Lastreopsis, Arachniodes, Tectaria,* and *Bolbitis.* If this extrapolation is extended it would indicate that on this basis the onocleoid ferns are allied with *Athyrium, Cystopteris, Diplazium, Woodsia,* and the thelypteroids. The structure of the vascular strand does not indicate any relation between the onocleoids and the blechnoids.

Trichomes and scales.—The occurrence and structure of trichomes and scales varies within the onocleoid genera. Stipes of *Onoclea* are mostly glabrous except for occasional deltoid lanceolate, concolorous, buff-tan, entire scales, up to 4 mm long (fig. 9, *h*). The rachis of the sterile frond appears glabrous but is infrequently provided with a few linear-lanceolate scales 0.8 to 2.5 mm long and whitish trichomes (fig. 9, *a, d, e*). The scales are deltoid to deltoid-ovate and attenuate when fully mature. Their margins are more or less entire but with occasional projections tipped with globose reddish or brownish glandular cells. There is a complete morphological sequence from unicellular trichomes to large scales present on the rachis. The unicellular trichomes are about 200 microns long. Bi- to tricellular, biseriate scales occasionally occur. These are about 250 microns long and show the characteristic red-glandular terminal globose cell found frequently along the margins of mature scales.

The adaxial surface of fronds of *Onoclea* are provided with numerous small scales up to 1.4 mm long, and uni-, bi-, or multiseriate, linear scales 150 to 1000 microns long, with clear lumina and thin cell walls (fig. 9, *c, f, g*). Some of these scales possess glandular terminal cells. These trichomes and scales occur primarily along the veins and the midrib. The abaxial surface bears frequent very small unicellular trichomes along the midrib of the pinnae. These vary in length from 45 to 120 microns. The lower vein axils and rachis have scattered linear- lanceolate, attenuate scales 0.8 to 2.3 mm long and 200 microns wide at the base of the broadest. Some of the trichomes and small linear scales appear branched at the base (fig. 9, *i*).

The stipe base of *Onocleopsis hintonii* is profusely covered with linear-lanceolate to ovate-lanceolate, attenuate, concolorous, tan to reddish-brown scales (fig. 9, *k, p, q*). These are unbranched and appear to be nonglandular. They vary from 45 to 120 microns. The lower vein axils and rachis have scattered linear-lanceo-toward the apex of the stipe. Some of the lower scales possess a slightly darker median area caused by both thickening and deeper color in the cell walls of this region. The adaxial surface of the frond is mostly glabrous except for a few scattered whitish trichomes up to 240 microns long on the veins. The abaxial surface of the pinnae is covered with numerous small whitish trichomes along the basal portions of the veins and in their axils (fig. 9, *k–o*). These trichomes are linear to more or less clavate, two- or three-celled, and about 150 to 200 microns long and 40 microns wide. They appear to be nucleate and only slightly glandular at the apex. The epidermal cell at the base of the trichomes is frequently enlarged. At the base of the pinnae of the sterile frond are a few small linear-attenuate to lanceolate-attenuate scales, about 5 mm long and 0.7 mm wide. These are nonglandular, unbranched, and concolorous tan in color. Linear-lanceolate attenuate scales also occur infrequently along the lower portions of the rachis. Whitish, uniseriate trichomes with six to seven cells, 120 microns to 1 mm in length, are found throughout the rachis.

The stipe of *Matteuccia struthiopteris* is densely covered at the base with ovate to lanceolate, concolorous or dark-based scales, identical in overall structure to those of *Onocleopsis*. These scales vary in length from 5 to 10 mm and in width from 1.5 to 3 mm. Above the base of the stipe are abundant small uniseriate trichomes 70 to 200 microns long; these gradually diminish in number toward the apex. The structure of these trichomes is similar to those of *Onocleopsis*. Scales are mostly lacking on the upper portion of stipe and rachis. On the abaxial side of the blade of the sterile frond are linear-lanceolate to lanceolate, concolorous scales, 0.9 to 2.2 mm long and 0.4 mm wide. These scales have entire margins and are mostly nonglandular. Uniseriate trichomes one to seven cells long and up to 1.5 mm long occur on the main veins. At the point of attachment of the pinnae to the rachis are found a few linear scales about 0.7 mm long. Gland-tipped trichomes 175 to 230 microns long are scattered on the abaxial side of the sterile pinnae along the veins. The terminal cell of these glandular trichomes is reddish-brown and greatly enlarged. The adaxial side of the frond is mostly glabrous.

Scales found on the sterile stipe and frond of *Matteuccia intermedia* vary only slightly from those found in *M. struthiopteris* and *Onocleopsis*. The stipe scales abundant near the base are ovate-lanceolate, 0.8 to 1.3 cm long and 3 to 4 mm wide. Rachis scales are lanceolate and entire, or with slightly toothed margins (fig. 9, *z*). These scales are 1.6 to 8.0 mm long and 0.2 to 1.5 mm wide. The adaxial surface of the frond is glabrous except for an occasional linear, brown scale about 1 mm long along the edge of the pinna midrib. The abaxial surface of the pinnae is covered with linear, uniseriate, multicellular, brownish trichomes and scales 0.5 to 1.7 mm long and about 35 microns wide. On some pinnae these are laneolate-attenuate, about 5 mm long and 0.5 mm wide at the base. The color of the scales varies from concolorous tan, to tan with a broad dark median strip, to tan with dark tips, or tan with dark bases. The color pattern reflects the coloring of the cell walls of the scales. Most specimens from Sikkim and India have concolorous scales; specimens from Tibet have scales with dark bases; and specimens from China exhibit scales with dark median strips, darkened tips or bases.

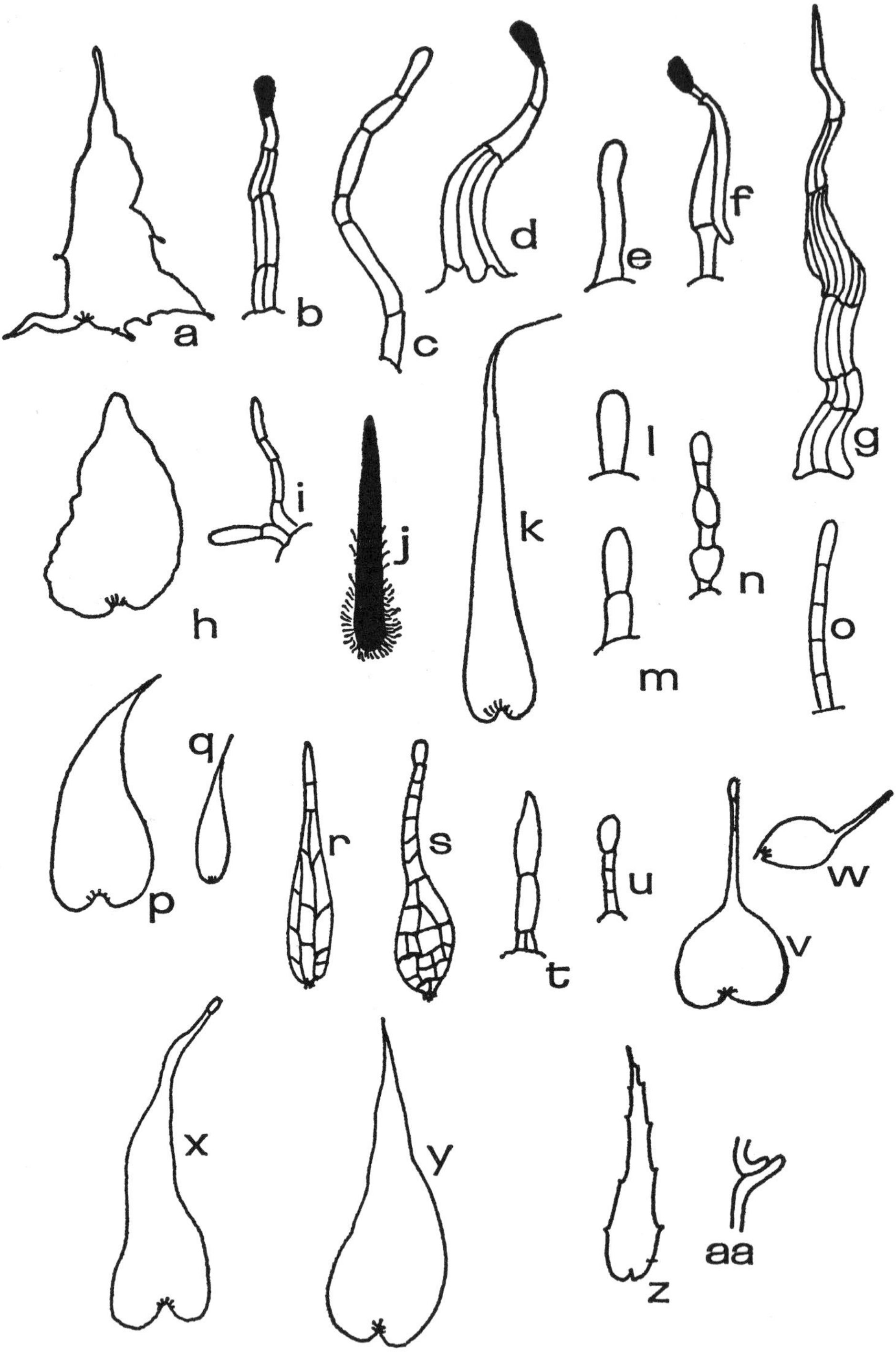

Fig. 9. Trichomes and scales. *a–j, Onoclea sensibilis. a,* rachis scale, × 16; *b,* rachis scale, × 80; *c,* trichome, adaxial frond surface, × 60; *d,* rachis scale, × 90; *e,* rachis trichome, × 90; *f,* scale, adaxial frond surface, × 65; *g,* scale, adaxial frond surface, × 48; *h,* stipe scale, × 7; *i,* scale, abaxial frond surface, × 45; *j,* rhizome scale, × 3. *k–q, Onocleopsis hintonii. k,* scale, × 12; *l,*

The stipe and rachis of the sterile frond of *Matteuccia orientalis* are profusely covered with scales, gradually diminishing in size toward the apex. Scales at the base of the stipe are ovate-lanceolate to lanceolate, attenuate, 0.7 to 2.5 cm long and up to 5 mm wide (fig. 9, *x, y*). They are unbranched and nonglandular and are similar in structure to those of other species of *Matteuccia* and *Onocleopsis*. The scales are concolorous tan.

The adaxial surface of the sterile frond has a few uniseriate, multicellular, trichomes scattered on the veins (fig. 9, *t, u*). These have dark cell walls, clear lumina, and are 240 to 600 microns long and nonglandular. On the abaxial surface of the frond the veins are commonly covered with uniseriate, multicellular trichomes about 0.2 to 0.6 mm in length. The rachis and the midribs of the pinnae bear a few linear-lanceolate scales 1.4 to 6 mm long (fig. 9, *r, s*), and frequent saccate-attenuate scales with an enlarged reddish glandular terminal cell (fig. 9, *v, w*). These latter scales are 1.5 to 3 mm long and are not found in any other species of the onocleoid complex.

Comparisons with other genera.—It has long been believed that the primitive dermal appendage in ferns was the simple trichome composed of a single row of cells (Bower, 1914). The flattened scale, broad at the base and of varying form, is thought to be derived and is characteristic of more advanced ferns. The ontogeny of the scale indicates that in many cases it originates as a simple unicellular trichome, and passes through several more complex stages, finally terminating in the mature scale. In the onocleoid ferns it is easy to interpret the simpler trichomes as representing only ontogenetic stages of the broader scales. Also, in the onocleoid ferns as well as in many Aspidiaceae, Blechnaceae, and other advanced groups, broad scales cover the rhizome and stipe bases, diminishing in size and frequency toward the apex of the frond. In many species the frond apex is completely glabrous.

The condition of having simple trichomes terminated by a glandular cell and broader scales with gland-tipped branches and apices occurs not only in onocleoid ferns but also throughout the genus *Blechnum* and also in species of *Brainea, Stenochlaena, Doodia,* and *Woodwardia* (Bower, 1914). These structures are present also in *Asplenium* and *Scolopendrium* (Bower, 1914), *Nephrolepis cordifolia, Athyrium, Thelypteris, Dryopteris, Cystopteris, Woodsia, Dennstaedtia,* and in many other genera of aspidioid affinities. Indeed, distribution of glandular trichomes and scales is so broad that its usefulness as a comparative character is inconclusive.

Of those scale characters that have been investigated, it is apparent that only overall shape and structure of cell walls can be used comparatively. Within Blechnaceae, the common type of scale is stiff, erect, rather pointedly tipped with an elongate acicular cell, and with narrow, elongated cells with thick, heavy dark walls. This type of scale has been observed on several species of *Blechnum* (*B. occidentale, B. gibbum, B. lehmanii, B. ensiforme*), *Brainea insignis,* and *Doodia media.* It occurs also in *Athyrium crenatum* and *Dryopteris varia.* Scales

trichome, × 90; *m,* trichome, × 100; *n,* trichome, × 65; *o,* trichome, × 70; *p,* scale, × 4; *q,* scale, × 6. *r–y, Matteuccia orientalis. r,* scale, × 20; *s,* scale, × 26; *t, u,* trichomes, × 65; *v, w,* scales, × 15; *x,* scale, × 27; *y,* scale, × 7. *z–aa, Matteuccia intermedia. z,* scale, × 5; *aa,* scale-tooth, × 150.

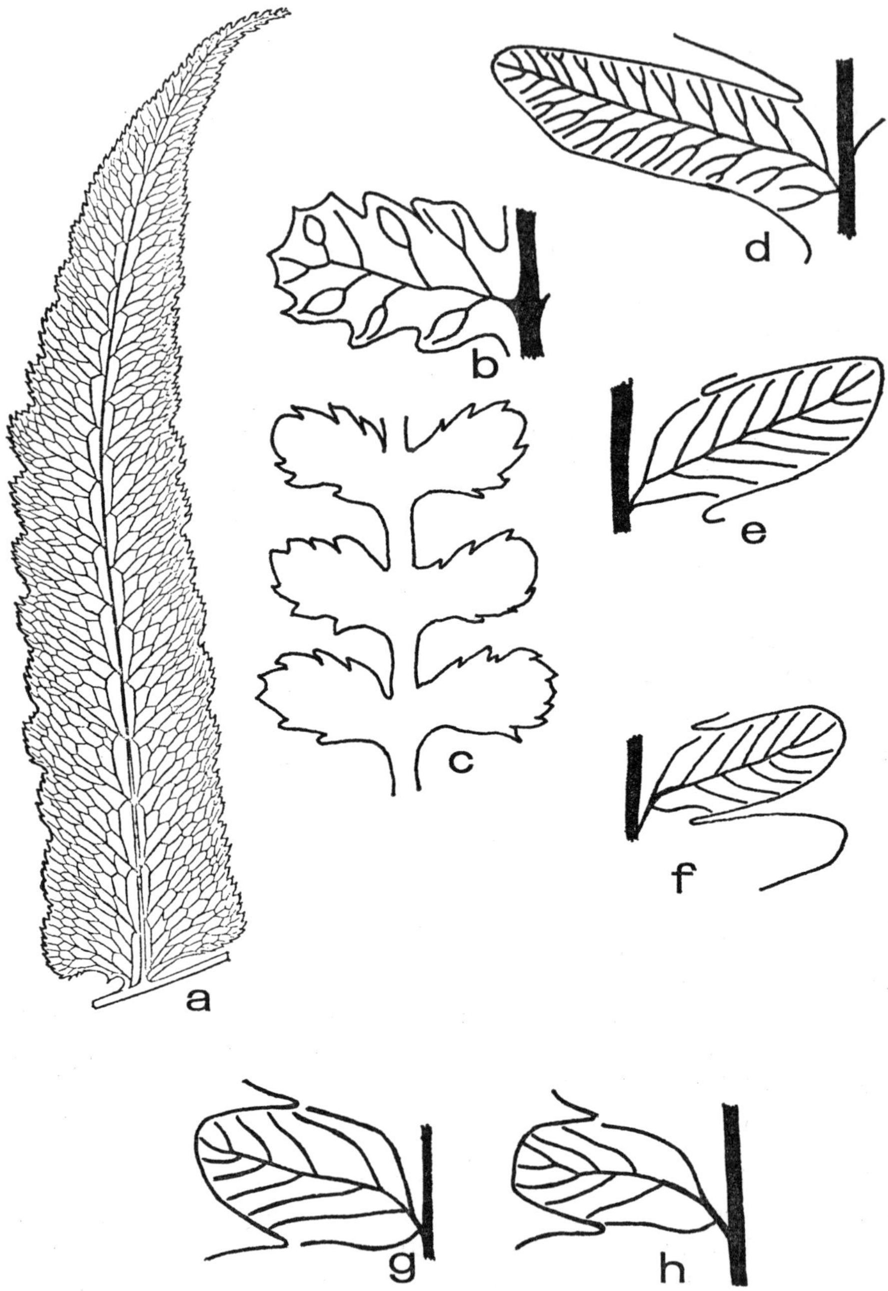

Fig. 10. Venation of sterile fronds. *a–c, Onocleopsis hintonii. a,* normal sterile pinna, × .8 (reprinted from F. Ballard, 1948. Further notes on *Onocleopsis.* Am. Fern J. 38:125–132; by permission) *b,* pinna lobe intermediate between sterile and fertile, × 5; *c,* pinna intermediate between sterile and fertile, × 2.5; *d–e, Matteuccia orientalis. d,* pinna lobe, × 2; *e,* pinna lobe, × 3; *f, M. struthiopteris,* pinna lobe, × 4; *g–h, M. intermedia,* pinna lobe, × 5.

similar to those of the onocleoid genera have been found also in *Athyrium nikkoense, A. nakanoi, Cystopteris setosa, Diplazium aspidioides, D. kominatoensis, Polystichum munitum, Woodsia manchuriensis,* and *Woodwardia areolata.*

It is apparent from the evidence at hand that the scales found in the onocleoid ferns are most similar to those of various genera of aspidiaceous affinity. In contrast, they differ substantially from the scales found in Blechnaceae, with the exception of *Woodwardia areolata.*

Venation.—Onocleoid ferns have two patterns of venation in the pinnae. *Onoclea sensibilis* and *Onocleopsis hintonii* exhibit lonchopteroid venation, in which the secondary veins form a reticulate network including a long costal areole, with the pinnae lobes attached by the whole breadth of their base (fig. 10, *a*). *Matteuccia* exhibits a pecopteroid venation pattern, in which the secondary veins do not form a network but are free from a definite midrib to the margin; the pinnae lobes are small and attached by the whole breadth of their base (fig. 10, *d–h*).

In *Onoclea* and *Onocleopsis* the long costal areoles arise at one pinna lobe midvein and run along the frond rachis to the next. In each pinna lobe in *Onoclea* there are four or five costal areoles along the midrib on either side. Veins in both species run at an obtuse angle (from the rachis) to the margin, anastomosing frequently, and producing lateral areoles. These areoles are longer and narrower in *Onoclea* than in *Onocleopsis.* The primary difference in venation pattern between the two genera lies in the presence or absence of a partially or completely formed vein connecting the vein endings along the margin of the pinnae. In *Onoclea* many of the vein endings turn apically at the margin and connect to the next vein terminus. This system is by no means complete although unusual epidermal tissue present at the margin makes it appear to be. Close examination for vascular elements shows their frequent absence from this region. In *Onocleopsis,* veins have never been observed to turn apically, although the tissue of the epidermal region appears to be similar to that found in *Onoclea.* The venation pattern in *Onocleopsis* appears to be constant in all of the known populations.

In the successive development of fertile fronds of *Onocleopsis,* fronds intermediate in structure usually appear from the apex shortly before production of the fertile fronds. These fronds are more or less deeply lobed with venation intermediate between the reticulate pattern of the sterile frond and the free pattern of the fertile frond (fig. 10, *b, c*). In these lobes veins arise from the midrib and run to the margin either freely or with terminal anastomoses, producing oval to circular areoles. Occasionally fronds are produced with a completely free venation pattern; in such cases, the overall venation and frond is similar to that of *Matteuccia.*

The venation pattern in all three species of *Matteuccia* is similar, varying only in the number of veins arising from the midrib of the pinna lobes. The lowermost veins arise from the midrib at the base, usually at an angle acute to the rachis and arcuate toward the margin. Higher veins arise at an angle obtuse to the rachis and proceed more or less directly to the margin.

The number of veins in the pinnae lobes of the three species is summarized in table 5.

Number of veins per pinna lobe is an excellent character by which to distinguish *M. intermedia* from *M. orientalis:* although the sterile frond of *M. intermedia* is similar in morphology to that of *M. struthiopteris,* the number of veins is far less. The relatively high number of vein endings in *M. orientalis* is the result of dichotomous branching of the lateral veins. This phenomenon is absent in the other two species, because of their smaller pinnae.

Comparisons with other genera.—Bower (1914) regarded the venation pattern of *Woodwardia areolata* and *Onoclea sensibilis* (which are similar) to be recently derived from the open-dichotomous pattern exhibited by the cyatheoid ferns and

TABLE 5

NUMBER OF VEINS IN PINNULES OF MATTEUCCIA

	Species		
	M. struthiopteris	*M. intermedia*	*M. orientalis*
No. veins above the midrib..........	6–11 ($\bar{x}$ = 8.3)	4–7 ($\bar{x}$ = 5)	7–22 ($\bar{x}$ = 9.7)
No. veins below the midrib..........	7–12 ($\bar{x}$ = 8.9)	4–8 ($\bar{x}$ = 5.4)	7–22 ($\bar{x}$ = 10.5)

Matteuccia. Investigation of intermediates between sterile and fertile fronds in *Onocleopsis* indicates how easily this could take place by progressive formation of tissue between the pinnules and the subsequent vein dichotomies and anastomoses. Hara (1964) has found a very close relationship between reticulate and open-dichotomous venation in *Onoclea* and *Matteuccia* and other ferns studied. He states that the venation pattern of *Onoclea* is reticulate because the forked branches of the dichotomies fuse and that there "may be no fundamental difference between the process of formation of forked procambia in open dichotomy and the process of forking of the procambium in *Onoclea*" (p. 386). He concludes that the principal developmental pattern of venation in *Onoclea* is dichotomous, as is true of most ferns. It is apparent from his study and from the occurrence of lonchopteroid venation patterns in widely separated genera of Aspidiaceae and Blechnaceae that the character may have no phylogenetic implications. In fact Bower (1923) stated that venation patterns "cannot be used as a consistent or trustworthy basis for the phyletic seriation of the Filicales at large. This conclusion . . . naturally follows from the fact now demonstrated for so many isolated genera that a closed venation has originated repeatedly in distinct stocks. It is in fact a widely homoplastic character, determined in great measure by physiological necessity or convenience" (p. 96).

I have seen venation patterns similar to that of *Onoclea* in *Woodwardia areolata* and *Arcypteris* spp. Although most Blechnaceae have a long costal areole on either side of the midrib, the remaining veins run freely to the margin, and the venation pattern of nearly all species of *Blechnum* is free.

Epidermal-cell pattern.—Wagner (personal communication) has stated that

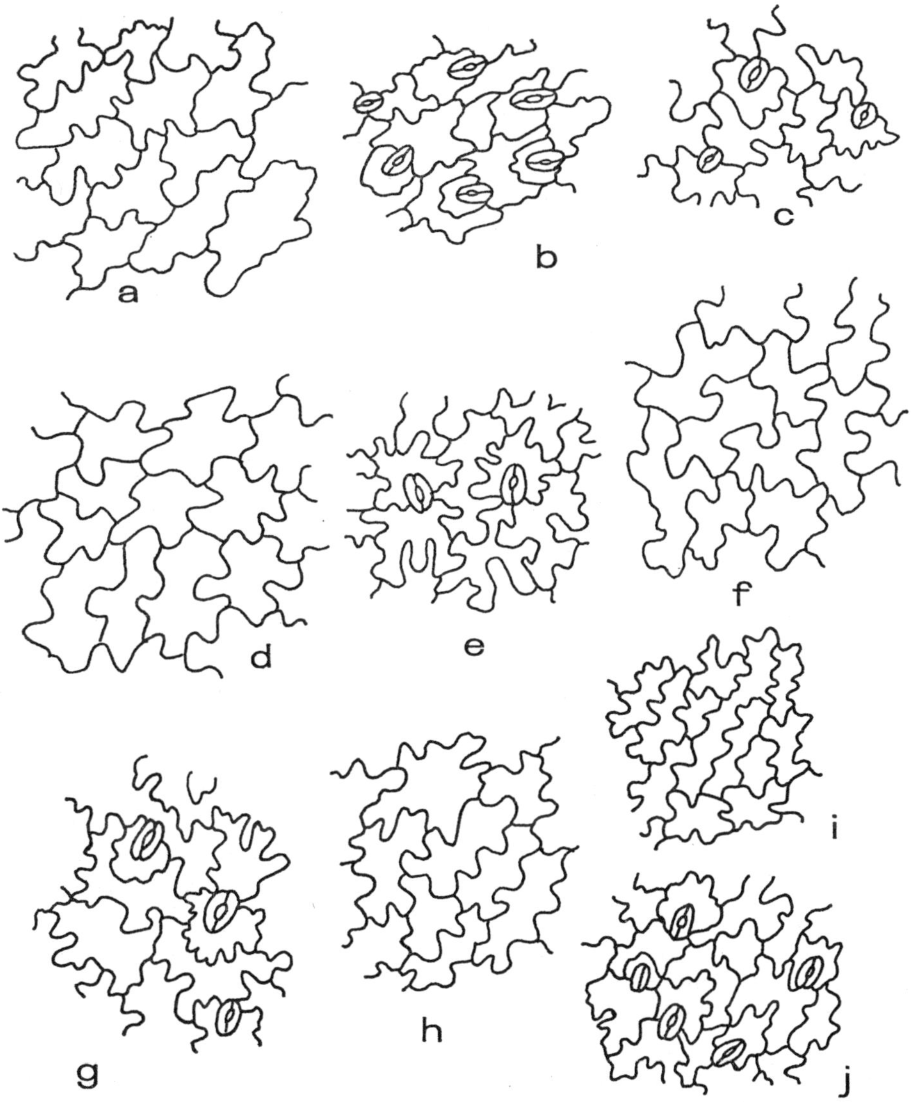

Fig. 11. Epidermal cell patterns of sterile frond. *Onoclea sensibilis: a,* upper, × 160; *b,* lower, × 140; *Onocleopsis hintonii: c,* lower, × 130; *d,* upper; *Matteuccia struthiopteris: e,* lower, × 165; *f,* upper, × 175; *M. orientalis: g,* lower, × 175; *h,* upper, × 155; *M. intermedia: i,* upper, × 110; *j,* lower, × 130.

epidermal-cell patterns probably will offer much in the way of comparative evidence for phylogenetic speculation on relationships in the ferns. However, in his work on *Diellia* he could find no consistent trend or pattern within the assemblages involved. Cell pattern and stomata were observed in the onocleoid ferns and a few postulated relatives. In general, the patterns exhibited by *Onoclea, Matteuccia,* and *Onocleopsis* are remarkably similar (fig. 11, *a–j*). Of the species of *Matteuccia, M. intermedia* exhibited the most diverse pattern, a result primarily of the length of the cells on the abaxial surface. Stomatal sizes in these three genera are summarized in table 6.

The epidermal-cell pattern exhibited by the onocleoids differs from those patterns of *Diellia* and *Nephrolepis* (Wagner, 1952). Patterns shown by the species of *Blechnum* observed appeared to differ primarily in the length of the cells on the adaxial surface. This was true also for *Peranema cyatheoides,* which had much longer cells, and *Diacalpe aspidioides,* which had shorter, less undulate cells. The pattern of cells surrounding the stomata on the abaxial surface was similar in all species examined.

TABLE 6

STOMATA SIZE IN ONOCLEOID FERNS
(In microns)

Species	Length	Mean	Width	Mean
Onoclea sensibilis	34–43	36	20–35	25
Onocleopsis hintonii	47–55	51	35–38	37
Matteuccia struthiopteris	44–63	51	25–40	32
M. intermedia	48–60	55	35–44	37
M. orientalis	50–63	57	33–38	35

Anatomy of the blade.—Midrib: Reduction in the diameter of the midrib in relation to that of the stipe results in fusion of the two distinct vascular strands present at the base of the stipe. In all three genera of onocleoid ferns, these fuse into a more or less U-shaped single strand. In *Onoclea* the midrib itself loses most of its adaxial groove and appears terete. The vascular stele is surrounded by a cortex of more or less uniform parenchyma cells, all with thin walls except for the sclerified outer cell layers. In *Onoclea* this region is four to seven cells thick and has secondary cell walls but no lignification. In *Onocleopsis* these cells are lignified and are about four cell layers in thickness. The midrib of *Matteuccia struthiopteris* is strongly grooved on the adaxial side. The outer cortex of the abaxial side of the midrib consists of five or six layers of cells with thickened, lignified walls. The midrib of *Matteuccia orientalis* shows a small groove on the abaxial surface. The outer cortex surrounding the stele is two to four cell layers thick and consists of thick-walled, partly lignified cells. The interior pith is composed of large, thin-walled parenchyma.

Frond blade: The tissue of the frond blade in all three genera is described here as bifacial with a palisade and a spongy layer, although in ferns that I have observed (Lloyd, 1965) it is difficult to separate the two regions because the size and shape of mesophyll cells are similar. Generally, the palisade layer refers to that region of closely packed cells without great amounts of air space and the spongy layer to the more loosely organized region of the blade. These two regions correspond in position to the palisade and spongy regions of angiosperm leaves. Table 7 summarizes most of the salient features of frond anatomy in onocleoid ferns.

Tissue comparisons in ferns have yielded little in the way of comparative characteristics useful for phylogenetic purposes. Wagner (1952) compared sev-

TABLE 7

SALIENT FEATURES OF FROND ANATOMY OF THE ONOCLEOID FERNS

Character	*Onoclea s.*	*Onocleopsis h.*	*M. struthiopteris*	*M. orientalis*
Vascular bundle of midrib				
Width (microns)	230	170	140	130
Length (microns)	500	...	...	480
Frond thickness (microns)	80–180	130–300	110–200	150–330
Cell layers in palisade	3–5	...	...	1–4
Percent of frond thickness in spongy layer	33–50	50–66	33–50	50–66
Percent air space in spongy layer	33	55	35	60

eral species of *Diellia* with genera of Davalliaceae and Aspleniaceae and found the comparisons to be inconclusive.

In comparing the anatomy of the onocleoids with that of genera of Blechnaceae, two major differences have been noted. The first is the presence of a sclerified sheath around the vascular bundle of the rachis in members of Blechnaceae. This is similar to the bundle schlerenchyma reported for the stipe bundles in these species. The onocleoid genera do not have this sclerified region around the vascular bundle. The second major difference is the presence of a very densely packed palisade layer with a dense protoplasm that stains gray to purple. This condition was found in all species of *Blechnum* investigated, as well as in *Woodwardia areolata,* although the intensity of color in the last species was less than that found in *Blechnum.* Of the species of *Aspidiaceae* investigated, only *Cyrtomium falcatum* approached the palisade condition found in Blechnaceae. In the onocleoid genera, there is no apparent difference in cell structure and staining between spongy and palisade layers; in this feature *Onoclea, Onocleopsis,* and *Matteuccia* must be considered to fit into the general pattern prevalent in Aspidiaceae.

FERTILE FROND

The fertile frond of all three genera of onocleoid ferns is highly reduced in comparison to the sterile frond. It is shorter, erect, brown to black when mature, with sori enclosed both by an indusium (except in *M. intermedia*) and the overrolled lamina of the frond. Fertile fronds of *Onoclea* are produced during the summer growing season and into early fall. They are lanceolate-oblong, bipinnate with the pinnules rolled up into globular divisions about 4 mm long and 3 mm wide. The fertile fronds of *Onocleopsis* are produced from August to September. They are oblanceolate to elliptic, bipinnate with the pinnules rolled up into three to four globular divisions 1.5 to 3 mm long and 1.5 to 2 mm wide. The pinnae are usually curved at their ends toward the base of the plant. The basal pinnae are broader and sterile.

Fertile fronds of *Matteuccia struthiopteris* are narrowly oblanceolate with spreading-ascending linear moniliform pinnules 2 to 5 cm long and 2 to 4 mm wide. The basal pinnae are diminutive, some of them sterile and somewhat foliaceous. Fertile fronds of *M. orientalis* and *M. intermedia* differ from those of *M. struthiopteris* in being more or less truncate at the base with linear and non-moniliform pinnae. The adaxial side of the pinnae also is dark reddish brown in color.

Stipe anatomy.—The anatomy of the stipe is similar to that of the sterile frond in all three genera except for the presence of additional sclerification in the supportive cortical cells. This feature accounts for the erect persistence of the fertile fronds throughout the winter.

Trichomes and scales.—Structure of trichomes and scales found on the fertile fronds of onocleoid ferns is similar to those occurring on sterile fronds, but their distribution differs somewhat. The stipe of *Onoclea* possesses occasional broadly deltoid to lanceolate scales, which are about 2.5 mm long and 1.7 mm wide at their broadest. The scales are concolorous except for an occasional darkening at the point of attachment. The scales on the rachis are distributed like those of the sterile frond. Trichomes frequently are found, however, on the adaxial vein surfaces of the fertile pinnules.

The stipe base of *Onocleopsis* is profusely scaly. The rachis, upper stipe, and adaxial midribs of the pinnae are densely covered with uniseriate trichomes, 0.1 to 0.3 mm long. The abaxial surface of the fertile frond is glabrous.

The base of the fertile stipe in *M. struthiopteris* is abundantly scaly. The upper stipe possesses a few, scattered, whitish trichomes, which are more numerous above than below. The pinnae are mostly glabrous except for a few scattered trichomes on the abaxial surface. One population from British Columbia exhibits numerous glandular trichomes on the adaxial surface of the frond, both on the veins and on the intervening tissue.

The stipe bases of both *M. intermedia* and *M. orientalis* also are densely covered with scales. In *M. orientalis* the rachis has numerous linear scales that diminish toward the apex. The adaxial surface of the blade has occasional trichomes on the midrib, 235 to 250 microns long. In *M. intermedia* the abaxial midrib of the fertile frond infrequently possesses linear scales 1 to 2 mm long and 0.1 to 0.2 mm wide.

Venation.—The venation pattern of normal fertile pinnae in all genera of onocleoid ferns is free. In *Onoclea* each pinnule possesses a midrib-like vein with two to four more or less opposite emerging veins, these branching once or twice before reaching the margin (fig. 12, *a, b*). The fertile pinnae of *Onocleopsis* have three to five more or less opposite pinnules. The pinnae have a midrib with four to five opposite to alternate branches, each branch dichotomizing once or twice before reaching the margin (fig. 12, *c*). In pinnae which are not completely reduced and which have tissue along both sides of the midrib, there is usually a long, narrow, costal areole on either side.

Venation of fertile pinnae of *Matteuccia struthiopteris* is somewhat variable as to number of veins, depending on size of pinnules (fig. 12, *d–f*). Small, completely reduced pinnules will have a midvein with only three branch veins. These do not dichotomize before reaching the margin. The veins arise in exactly the same manner as in the sterile frond, with the lowermost veins curving toward the apex of the pinnule. Larger pinnules may have four to five branch veins.

Fronds intermediate between sterile and fertile fronds occur occasionally. In them the outer margin of the pinna lobe is reflexed only slightly; sori occur singly at the end of each vein, each with a membranous extrorse indusium attached to the vein on the midrib side of the sorus (fig. 12, *g, h*).

The venation of *M. orientalis* and *M. intermedia* is similar (fig. 12, *i–m*). In these species lateral veins arise regularly along both sides of the midrib of the pinna, and branch once to thrice before terminating about halfway to the margin. The marginal area of the pinnae of the fertile frond is composed of linear, thick-walled cells. Sporangia arise at the ends of the veins.

Indusium.—An indusium is present in all species of onocleoid ferns except *M. intermedia*. In *Matteuccia orientalis* the indusium is usually complete; in several specimens, however, it was only partially developed and in some it was difficult to find.

In *Onoclea* the orientation of the indusium is difficult to ascertain. In most cases it is extrorse, attached beneath the sorus at the base of the receptacle on the midrib side of the sorus. It is one or two cells thick and completely envelops the sorus. There are usually five to six sori in each pinnule, and it is difficult to determine from transections whether all indusia arise extrorsely.

The indusium in *Matteuccia struthiopteris* is similar to that of *Onoclea*, being membraneous and one or two cells in thickness. In fronds intermediate between sterile and fertile fronds, it appears to be saccate, arising from the vein in an extrorse manner. In completely reduced pinnules, it appears to be attached at the base of the receptacle, completely enveloping the sorus.

In *Onocleopsis*, the indusium is extrorse, arising from the base of the receptacle and completely enveloping the sorus (fig. 13, *a*). It is two or three cells thick at the base and one cell thick above.

The indusium in *Matteuccia orientalis* arises on the midvein side of the receptacle and envelops the sorus to varying degrees. It is membranous, two- to three-celled at the base and one-celled above (fig. 13, *b*).

Soral structure and frond anatomy.—The midrib structure of the fertile frond of onocleoid ferns is similar to that of the sterile fronds. In *Onoclea* all cells of the cortex have lignified secondary walls and the adaxial epidermal cells of the pinnae are highly cutinized. The mesophyll of the frond is reduced in quantity and is not differentiated into palisade and spongy tissues. Cutinization of the adaxial epidermis and lack of differentiation of mesophyll occur also in *Matteuccia*. In *Onocleopsis* there is no differentiation of mesophyll, although the adaxial epidermis is no more cutinized than in the sterile frond.

Sori in all three genera are produced at the ends of the free veins. When young, they are circular and discrete and only upon maturation do they merge into the sori surrounding them. This gives the impression of a linear sorus in *Matteuccia*. There is a situation parallel to this in *Elaphoglossum* which has an acrostichoid-like sorus. In this case, the sori are distinct and circular when immature but later merge together over the entire abaxial surface of the frond, giving an acrostichoid-like appearance (Lloyd, 1970).

Development of sporangia within each sorus in onocleoid ferns is more or less simultaneous, except that in younger stages it was observed that some sporangia

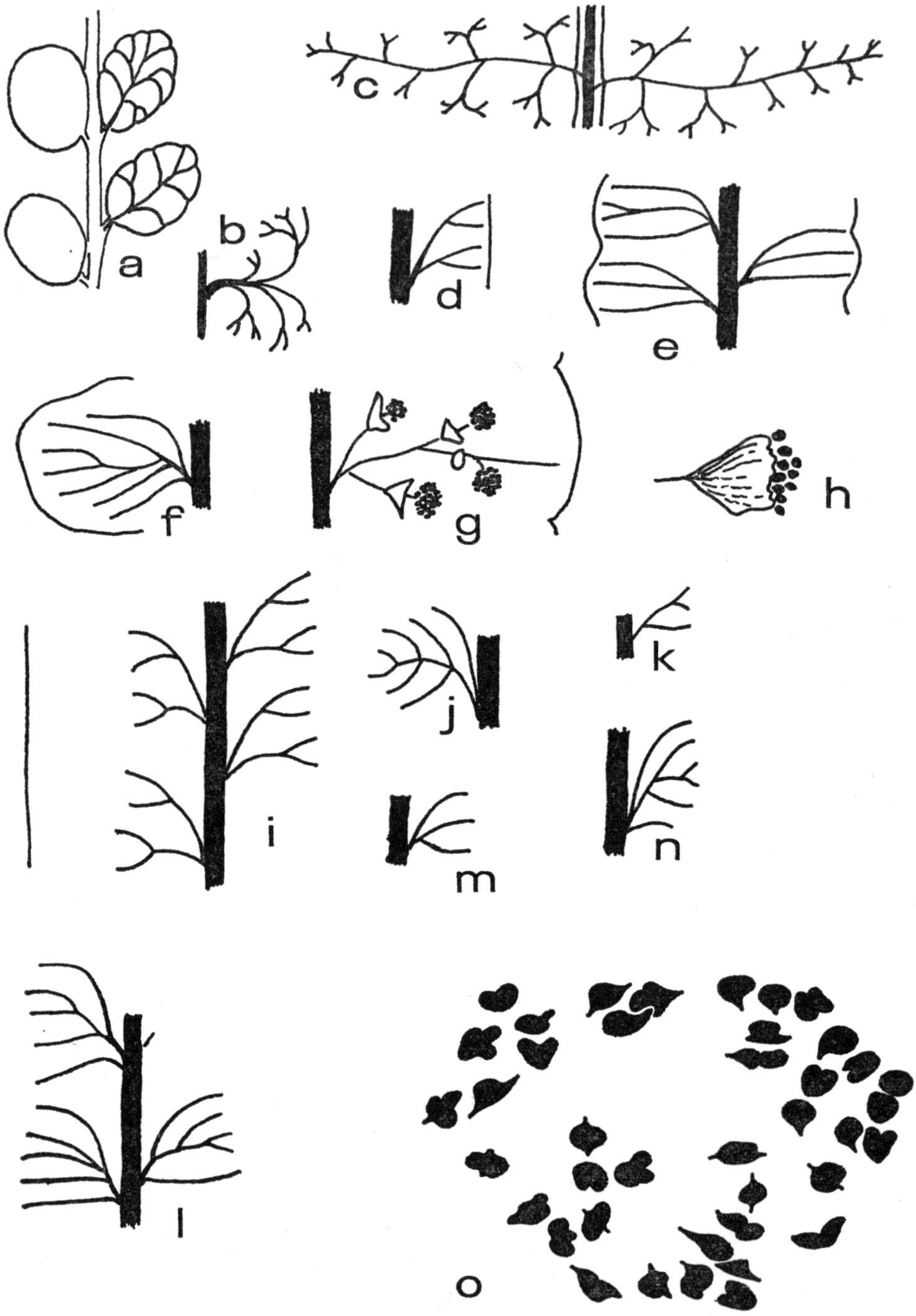

Fig. 12. Venation of fertile pinnae; chromosomes. *a–b Onoclea sensibilis: a,* pinnae, × 5; *b,* venation, × 5. *c, Onocleopsis hintonii,* venation of pinnae, × 4. *d–h, Matteuccia struthiopteris: d, e,* venation, × 5; *f,* × 10; *g,* pinnule intermediate between sterile and fertile with sori at tips of veins and extrorse indusium, × 7; *h,* sorus of *g. i–k, Matteuccia orientalis. i,* venation, × 3; *j, k,* × 7. *l–n, Matteuccia intermedia. l,* × 1.4; *m,* × 7; *n,* × 4.5. *o, Onocleopsis hintonii,* chromosomes, n = 40.

are produced later than others. This sequence, by strict definition, is gradate, although all sporangia are mature at the time of frond dehiscence. Development of sori on the frond proceeds gradately in an acropetal direction from the base. Here again, the sori are mature on the entire frond before dehiscence takes place.

The sporangia are typical for most advanced ferns. The major variable is number of indurated annulus cells. Table 2 summarizes these numbers in the onocleoids and the other ferns investigated.

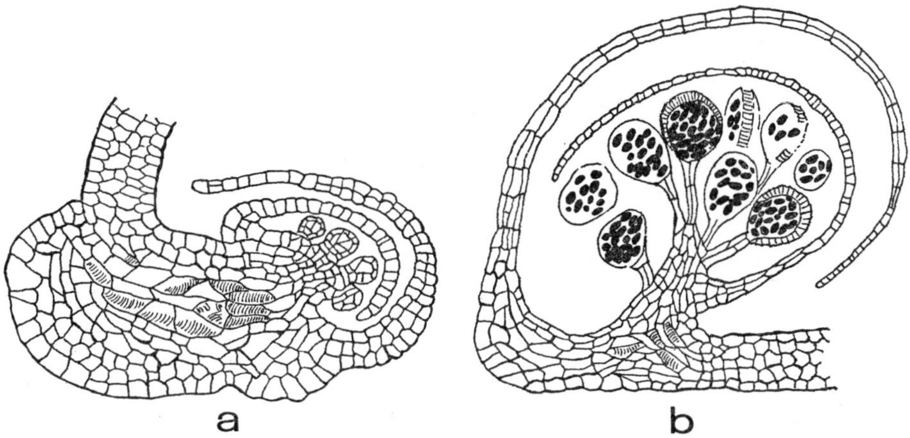

Fig. 13. Soral structure and indusia. *a, Onocleopsis hintonii,* transverse section of immature sorus, showing young sporangia and extrorse indusium; *b, Matteuccia orientalis,* transverse section of mature sorus.

The number of indurated annulus cells in the onocleoids varies from 21 to 45 ($\bar{x} = 28$). *Onocleopsis* has the smallest number, 21 to 26 ($\bar{x} = 24$) and *Matteuccia orientalis* the largest, 27 to 45 ($\bar{x} = 35$). *Matteuccia intermedia* has 28 to 36 ($\bar{x} = 32$). *Onoclea sensibilis* has 22 to 41 ($\bar{x} = 28$) annulus cells.

Number of indurated cells in species of Aspidiaceae examined varies from 10 to 29 ($\bar{x} = 15$). The largest number occurs in *Woodsia,* 10 to 29 ($\bar{x} = 17$). In Blechnaceae, the largest number is found in *Woodwardia,* 19 to 27 ($\bar{x} = 22$). It is apparent from these data that none of the other species investigated equals onocleoid ferns in number of indurated annulus cells, and a correlation with either Blechnaceae or Aspidiaceae cannot be made.

Of further interest is the fact that *Matteuccia intermedia* possesses a number of indurated cells somewhat intermediate between those of *M. struthiopteris* and *M. orientalis,* but the number is more similar to that of *M. orientalis,* as is the structure of the entire fertile frond.

Comparison of the fertile frond with other genera.—Dimorphism: Bower (1914) questioned whether presence of heterophylly was a "trustworthy" characteristic for phylogenetic comparisons between ferns. He then went on to utilize the presence of dimorphic fronds in the onocleoid ferns as "a valid point establishing their relationship with *Blechnum* . . ." (p. 417). He also suggested that this feature is a "physiological disability" and that it represents the primitive state in Blechnaceae. A close examination of occurrence of dimorphic fronds in

Filicales shows that it manifests itself in several unrelated families. These include Osmundaceae (*Osmunda*), Anemiaceae (*Anemia*), Plagiogyriaceae (*Plagiogyria*). Adiantaceae (*Anopteris, Cryptogramma, Trachypteris*), Parkeriaceae (*Ceratopteris*), Davalliaceae (*Humata*), Polypodiaceae (at least 12 genera including *Christiopteris, Marginariopsis, Microgramma, Pycnoloma,* and *Oleandropsis*), Blechnaceae (*Blechnum* spp., *Stenochlaena, Woodwardia areolata*), and Aspidiaceae (at least 21 genera including *Elaphoglossum, Microstaphyla, Egenolfia, Polybotrya, Maxonia, Lomariopsis, Bolbitis, Cyclodium, Faydenia,* and *Quercifilix*). It is apparent that this character alone cannot be used to determine phylogenetic relationships.

Sorus and indusium: Position of the sorus, maturation of sporangia, and orientation and structure of the indusium have long been used in construction of phylogenetic systems. It is unnecessary here to summarize or even to discuss many of the systems which have been proposed or the presumed evolution of these characteristics. Of foremost importance, however, is their presence or absence in Blechnaceae and Aspidiaceae.

Bower (1914, 1928) held that onocleoid ferns form the primitive stock from which the blechnoids arose, not only because of the presence of dimorphic fronds in the two groups but also because of the nonindusiate condition found in *Matteuccia intermedia*. He believed that *M. intermedia* is the living fern most similar to the hypothetical ancestor of Blechnaceae. Bower also stated that if the rows of sori of *Matteuccia* were linked together into a continuous sorus, together with formation of a commissural vein linking the veins beneath the receptacle, the coenosorus of *Blechnum* would result. The major objection to this argument is that the ontogeny of the two types of sori and their venation is completely different. In *Matteuccia,* the sorus is produced terminally on a vein; in *Blechnum* it is on top of a commissural vein linking two other veins. In *Matteuccia* the sori are initially circular, merging into a continuous, linear, collective sorus (but separated by indusia in *M. struthiopteris* and *M. orientalis*) only at maturity. In *Blechnum* the sorus is linear from the beginning, with sporangia arising all along the commissural veins. Although Bower thought that *M. intermedia* was primitive for the group evidence indicates that it is morphologically intermediate between *M. orientalis* and *M. struthiopteris* and probably of hybrid origin. Thus, the nonindusiate condition of *M. intermedia* must be considered to be recently derived.

Table 8 compares salient features of the fertile fronds of the onocleoid and the blechnoid ferns.

Another feature in which the onocleoids differ from the blechnoids is in the presence of a true membranous indusium in the former. No Blechnaceae possess a true indusium. The so-called indusium in such genera as *Woodwardia* and *Doodia* is fleshy, three to four cell layers thick at the base and tapering gradually into a coarse one- or two-layered apical region. It is always introrse, arising on the side of the sorus toward the margin and opening on the side toward the midrib. From its structure it is reasonable to suppose that it represents phylogenetically the inrolled pinna margin, which, in most species, has been modified by production of a flange and disintegration of the coenosorus into discrete elements. The indusium of the onocleoids, like that of many Aspidiaceae, is membranous, only one- to two-cell layers thick, and extrorsely oriented.

On the basis of soral position and indusial structure the only characteristics the onocleoids have in common with Blechnaceae are the gradate sorus and the basipetal maturation of sporangia within each individual sorus. These two characteristics are frequently found also in Aspidiaceae. Thus, there is no apparent reason to consider the onocleoids as being either closely related to or ancestral to the Blechnaceae.

TABLE 8

COMPARISONS OF THE FEATURES OF THE FERTILE FRONDS OF ONOCLEOID AND BLECHNOID FERNS

Character	Onocleoids	Blechnaceae
Margin	Recurved	Recurved (with or without flange)
Indusium	True—extrorse	False-recurved margin—introrse
Maturation of sporangia	Gradate	Gradate
Sorus location	Tip of veins	Top of commisural vein
Sorus shape initially	Circular	Linear
Sorus shape when mature	Linear (by fusion)	Linear
Mesophyll of blade	Not differentiated	Differentiated
Venation	Free	Reticulate

CYTOLOGY

Cytological studies were undertaken to ascertain the chromosome number of *Onocleopsis hintonii* and *Matteuccia orientalis* as well as to verify numbers previously reported for the other species of the onocleoid complex.

METHODS

Very young fertile fronds were fixed in Carnoy's solution (3 absolute Ethanol: 1 Glacial Acetic Acid) and were subsequently stored at low temperature. Young sporangia were macerated in acetocarmine with an iron needle and squashed after addition of the cover slip. The slide was then heated at one corner of the cover slip until vigorous boiling occurred. Utilization of both squashing and boiling assured good preparations and the release of spore mother cells from the sporangia. After observation of chromosomes the slides were made permanent by carefully removing the cover slip and adding a drop of Hoyer's medium (Beeks, 1955). Material made permanent in this manner has been found to retain its natural morphological structure for at least eight years, accompanied by a gradual clearing of the cytoplasm, facilitating observation of chromosomes.

RESULTS

Onoclea sensibilis exhibits 37 bivalents at metaphase I. There is no known variation in this number. There are reports of $n = 32$ (Gregory, 1904) and $2n = 64$ (Litardiere, 1921) for *Onoclea*. However, these counts were made early in this century and are probably inaccurate.

Sporocytes of *Onocleopsis hintonii* showed 40 bivalents at meiotic metaphase I

TABLE 9

KNOWN CHROMOSOME NUMBERS OF ONOCLEOID FERNS
(From Chiarugi, 1960; Fabbri 1963, 1969)

Taxon	Number	Source
Onoclea sensibilis	2n = 64	DeLitardiere, 1921
	2n = 74	Love & Love, 1961
	n = 32	Gregory, 1904
	n = 40	Friebel, 1933
	n = 37	Britton, 1953
	n = 37	Wagner, 1955
	n = 37	Lloyd (*Knobloch A*, UC)
Onocleopsis hintonii.........	n = 40	Lloyd (No. *4315*, UC)
*Matteuccia struthiopteris**....	n = 40	Sorsa, 1962, 1958
	n = 40	Friebel, 1933
	n = 40	Okuno, 1936
	n = 40	Mitui, 1965
	n = ca. 40	Britton, 1953
Matteuccia orientalis........	n = 40	Loyal in Mehra, 1961
	n = 40	Kurita, 1961
	n = 40	Friebel, 1933
	n = 40	Okuno, 1936
	n = 80	Lloyd (No. *4317*, UC)

* Recently, Fabbri and Menicanti (1970) reported n = 39 in cultivated plants of this species (see Caryologia 23:673–676).

(fig. 11, *o*). Counts of plants of *Matteuccia orientalis* from Japan indicate that these are tetraploid, with 80 bivalents at metaphase I. Previous counts of *M. orientalis* from plants of the Asian mainland have all been diploid ($n = 40$).

Table 9 summarizes all known counts for the onocleoid ferns.

DISCUSSION

The onocleoid ferns are characterized by two base numbers, $x = 37$ in *Onoclea* and $x = 40$ in *Onocleopsis* and *Matteuccia*. The presence of apparent aneuploidy is a phenomena not unknown in other ferns. For example, *Blechnum* is characterized by base numbers varying from 28 to 37; *Woodsia* has numbers of 38, 39, and 41; and in *Thelypteris* base numbers range from 27 to 36 (Chiarugi, 1960; Fabbri, 1963, 1965).

The base number of 40, found in *Onocleopsis* and *Matteuccia,* is rare in Aspidiaceae and has not been reported for Blechnaceae. It has been reported in *Athyrium, Polystichum, Tectaria, Gymnocarpium* ($n = 80$), *Quercifilix* ($n = 80$), *Heterogonium,* and *Stenosemia.* Copeland (1947) stated that these genera are essentially all part of the same natural group. In Davalliaceae, $x = 40$ is very common, having been reported from six of the twelve genera recognized by Copeland (*Araiostegia, Davallia, Davalliodes, Humata, Nephrolepis,* and *Oleandra*). It has also been noted in *Asplenium unilaterale* (Aspleniaceae), *Lindsaya prolongata* (Lindsaeaceae), and *Ceratopteris* (Parkeriaceae).

The base number of 37 is even rarer than 40 in indusiate ferns. It has been

TABLE 10

BASE CHROMOSOME NUMBERS OF SOME GENERA OF INDUSIATE FERNS, REPORTED SINCE 1940
(From Chiarugi, 1960; Fabbri, 1963, 1965)

Taxon	*X* Number
ASPIDIACEAE	
Athyrium	40, 41
Cyrtomium	41
Cystopteris	41
Diacalpe	41
Diplazium	41
Dryopteris	36, 41
Faydenia	32*
Thelypteris	27, 28, 30, 31, 32, 34, 35, 36
Peranema	41
Polystichum	40, 41
Tectaria	39, 40
Woodsia	33, 38, 39, 41
Gymnocarpium	80
BLECHNACEAE	
Blechnum	28, 29, 31, 32, 33, 34, ca. 37
Doodia	32
Stenochlaena	72, 73, 74
Woodwardia	ca. 32, 34, 35
ASPLENIACEAE	
Asplenium	36, 40
Camptosorus	35, 36
Ceterach	36
Loxoscaphe	ca. 70
Phyllitis	36
Pleurosorus	72
DAVALLIACEAE	
Araiostegia	40
Arthropteris	41, ca. 42
Davallia	40
Davallodes	40
Humata	40
Leucostegia	41
Nephrolepis	38, ca. 40, 41, 34–38
Oleandra	40, 41

* Count made in 1904 and the only one known for this genus.

reported only in *Onoclea* and in one species of *Blechnum* (Fabbri, 1965). *Stenochlaena* (Blechnaceae) has been found to have a haploid number of 74, considered to be a polyploid derivative of 37. The more common numbers in Blechnaceae vary from 28 to 35 (table 10). Of the nonindusiate ferns, Polypodiaceae and Grammitidaceae frequently have numbers based on 37.

The chromosomal data suggest that a relationship exists between *Matteuccia, Onocleopsis,* and the aspidioid ferns, in particular the *Polystichum-Tectaria-Athyrium* group. The relationship of *Onoclea,* on the other hand, is not clarified by these data. Although the base number of 37 is known for one species of *Blechnum* and unknown in Aspidiaceae other morphological criteria indicate that *Onoclea* is closely related to *Matteuccia* and *Onocleopsis*. Thus, the variation in base number among these genera is probably the result of aneuploidy.

CHROMATOGRAPHIC STUDIES

Methods and Materials

I have made an extensive investigation of the chromatographic patterns exhibited by the three onocleoid genera as well as those genera which have been suggested as relatives of the complex. Most of the material utilized in this study was obtained from herbarium specimens (see Appendix for citation of voucher specimens) or from plants in cultivation at the University of California Botanical Garden, Berkeley. The latter included material of *Onocleopsis hintonii, Onoclea sensibilis, Matteuccia orientalis* (tetraploid), and *Woodwardia areolata.* Dried material of these species was also chromatographed.

Material consisting of both sterile and fertile pinnae was macerated and compounds were extracted for 48 hours in from one to two ml of anhydrous methyl alcohol. Extracts were spotted on Whatman 3MM chromatographic paper. The chromatograms were run in two dimensions by the descending method. The solvent system for the first dimension consisted of t-butyl alcohol : glacial acetic acid : water (4:1:1, v/v). This procedure gave excellent separation and definition of spots. The dried chromatographs were analyzed using both UV light and ammonia vapor.

Results and Discussion

The onocleoid ferns are characterized by the occurrence of 29 compounds common to the three genera (table 12). Some of these (Nos. 1–5, 7, 13) are found frequently in chromatograms of most of the other species studied and thus have little value for comparative purposes. Other compounds are entirely restricted to the onocleoid complex, although they may occur in only one species (Nos. 14, 15, 17, 18, 20, 22, 24, 25, 27, 28). From chromatographic evidence, it is apparent that similarities in compounds do occur from group to group, even though there are marked differences. Table 13 summarizes the occurrence of compounds in this study, and Table 11 details their characteristics. It is significant that several spots are restricted to each taxon studied. For example, even though Aspidiaceae and Blechnaceae are thought to be closely related phylogenetically, they differ chromatographically in 46 spots, 24 of these being restricted to Aspidiaceae and 22 to Blechnaceae. However, chromatographic data at the family level are lacking, and little is known of the variation which occurs between complexes of families that have apparently long been evolutionarily distinct. It is perhaps even more significant that the families investigated have some similarities in spot patterns. For example, compounds 1, 2, 3, and 4 occur in some genera of all families studied. Compounds 5, 7, 8, 9, 11, and 21 occur in Aspidiaceae, Blechnaceae, and the onocleoid fern complex.

By calculating the total number of compounds observed in any two groups or genera in relation to the number of spots in common with the onocleoid ferns

TABLE 11
COMPOUND CHARACTERISTICS
(Of all compounds found)

Compound number	Rf Values		Color in UV	Color in UV + NH_4OH
	1-D	2-D		
1.	.51	.85	Blue	Light blue
2.	.51	.71	Blue	Yellow-green
3.	.51	.28	Purple	Yellow-orange
4.	.71	.43	Purple	Green
5.	.75	.43	Blue	Light blue
6.	.44	.82	. . .*	Light blue
7.	.33	.57	Salmon	Salmon
8.	.25	.21	Yellow	Yellow
9.	.33	.57	Blue	Light blue
10.	.83	.71	Blue	Light blue
11.	.42	.71	. . .	Dark blue
12.	.83	.50	Light purple	Light purple
13.	.75	.82	. . .	Dark blue
14.	.58	.25	Blue	Light blue
15.	.36	.50	Light blue	Light blue
16.	.25	.78	Light blue	Light blue
17.	.22	.43	Purple	Green
18.	.25	.07	Blue	Blue
19.	.58	.07	Yellow	Yellow
20.	.75	.28	Yellow	Yellow
21.	.80	.35	Blue	Blue
22.	.66	.21	Blue	Light blue
23.	.44	.78	. . .	Dark blue
24.	.71	.17	Purple	Purple
25.	.71	.32	White	White
26.	.83	.38	Blue	Light blue
27.	.66	.50	Light blue	Light blue
28.	.58	.64	. . .	Blue
29.	.36	.73	Blue	Light blue
30.	.53	.17	Purple	Orange
31.	.61	.46	Light yellow-green	Green
32.	.50	.53	Light orange	Light orange
33.	.42	.56	Yellow-green	Yellow-green
34.	.39	.64	Blue	Light blue
35.	.51	.82	Orange	Orange
36. = 1.				
37. = 2.				
38.	.71	.70	Yellow-green	Dark green
39.	.39	.10	Purple	Yellow
40.	.66	.21	Purple	Orange
41.	.80	.24	Purple	Purple
42.	.53	.50	Purple	Purple
43.	.28	.67	Purple	Brown
44.	.86	.63	Blue	Light blue
45.	.51	.35	Light blue	Lighter blue
46.	.61	.28	Blue	Light blue
47.	.69	.17	Blue green	Light blue-green
48.	.77	.28	Blue	Blue
49.	.80	.46	Yellow-green	Light yellow-green
50.	.35	.84	. . .	Purple

TABLE 11—*Continued*

Compound number	Rf Values		Color in UV	Color in UV + NH4OH
	1-D	2-D		
51.	.36	.19	Yellow	Yellow
52.	.86	.57	...	Dark purple
53.	.71	.28	Purple	Orange
54.	.66	.35	Purple	Purple
55.	.51	.27	...	Orange
56.	.50	.62	Purple	Brown
57.	.33	.70	Blue	Light blue
58.	.79	.38	Blue	Light blue
59.	.82	.62	...	Dark blue
60.	.20	.65	...	Light blue
61.	.76	.54	Purple	Orange
62.	.22	.17	Light orange	Yellow
63.	.36	.35	Dark orange	Yellow
64.	.30	.78	Orange	Light orange
65.	.19	.78	Blue	Blue
66.	.44	.85	Purple	Purple
67.	.42	.70	Purple	Brown
68.	.66	.57	Blue	Light blue
69.	.69	.70	Dark blue	Light blue
70.	.71	.78	...	Dark blue
71.	.86	.68	...	Dark blue
72.	.80	.50	Blue	Light blue
73.	.80	.35	Blue	Light blue
74.	.53	.50	Blue	Light blue
75.	.42	.57	Blue	Light blue
76.	.28	.57	Blue	Light blue
77.	.66	.88	Blue	Light blue
78.	.58	.57	Purple	Orange-brown
79.	.44	.53	Purple	Yellow
80.	.25	.47	Purple	Yellow
81.	.64	.70	Purple	Yellow
82.	.50	.73	Purple	Brown
83.	.19	.78	Purple	Purple
84.	.39	.69	Dark blue	Dark blue
85.	.75	.07	...	Yellow
86.	.47	.73	...	Violet
87.	.65	.84	Purple	Purple
88.	.82	.61	Purple	Brown
89.	.47	.73	Blue	Blue
90.	.50	.69	...	Purple
91.	.47	.50	Orange-brown	Brown
92.	.64	.58	...	Yellow
93.	.44	.73	Dark blue	Light blue
94.	.42	.28	Purple	Yellow
95.	.47	.17	Purple	Brown
96.	.39	.07	Purple	Orange
97.	.71	.61	...	Blue
98.	.38	.92	Light blue	Yellow-green
99.	.29	.76	Light blue	Yellow-green
100.	.41	.69	Purple	Purple

* Does not appear under ordinary UV light without NH4OH vapor.

TABLE 12

OCCURRENCE OF COMPOUNDS IN THE SPECIES STUDIED AND THEIR AFFINITY TO ONOCLEA SENSIBILIS AND THE ONOCLOID-COMPLEX

Taxon	Compounds present	Percent affinity to onocleoid complex	Percent affinity to Onoclea
Onoclea sensibilis	1, 2, 3, 4, 5, 6, 7, 8, 9, 10, 11, 23	41	100
Onocleopsis hintonii	1, 2, 4, 7, 8, 12, 13, 14, 15, 16, 17	37	28
Matteuccia struthiopteris	1, 2, 3, 4, 5, 7, 8, 9, 10, 11, 13, 18, 19, 20, 21, 22, 23	58	61
M. intermedia	1, 2, 3, 4, 5, 9, 11, 12, 22, 23, 25, 26, 29	44	47
M. orientalis (4x)	2, 3, 4, 5, 7, 9, 11, 12, 13, 22, 23, 24, 25, 26, 27, 28, 29	58	38
Athyrium nakanoi	2, 21, 26, 52, 58, 77, 78, 79, 80	9	5
A. nikkoense	52, 53, 58, 81, 82, 83, 84	0	0
A. strigillosum	26, 52, 81, 82	3	0
Cystopteris fragilis	1, 2, 3, 4, 13, 16, 19, 21, 52, 62, 78, 79, 85, 86	22	18
Diacalpe aspidioides	1, 3, 4, 29, 81, 87	13	21
Diplazium cristatum	2, 52, 58, 89	3	7
Dryopteris kominatoensis	1, 2, 3, 4, 13, 26, 52	20	28
D. lacera	1, 2, 4, 5, 13, 21, 39	20	28
D. sabaei	1, 2, 4, 5, 13, 52	17	30
Peranema cyatheoides	2, 3, 4, 13, 26, 39, 50, 52, 61	15	18
P. formosana	2, 3, 5, 7, 8, 13, 39, 50, 52, 61, 87, 88, 89, 90	16	25
Polystichum munitum	2, 3, 7, 13, 52, 82, 91, 92, 94, 95, 97, 98, 99, 100	10	14
P. munitum var. curtum	2, 3, 7, 9, 31, 39, 52, 82, 91, 92, 93, 94, 95, 96, 97	10	18
P. retroso-paleaceum	2, 5, 9, 13, 21, 52	17	21
Thelypteris normalis	2, 3, 4, 5, 8, 9, 10, 26, 52, 85	26	47
Woodsia ilvensis	1, 2, 3, 5, 13, 21, 56, 81	19	27
W. manchuriensis	2, 3, 4, 5, 7, 9, 11, 13, 23, 26, 81	33	57
Davallia mariesii	1, 2, 3, 4, 13, 62, 63, 64, 65, 66, 67, 68, 69, 70, 71, 72, 73, 74, 75, 76	11	15
Blechnum occidentale	1, 2, 5, 7, 8, 21, 45, 46, 47, 48, 49	18	29
B. sessifolia	1, 2, 3, 4, 5, 7, 12, 21, 31, 39, 40, 41, 42, 43, 44	22	30
Brainea insignis	2, 5, 7, 50, 52, 55, 56, 60, 61	8	18
Doodia caudata	1, 2, 5, 7, 9, 11, 49, 52, 56, 58, 59	9	37
Sadleria cyatheoides	1, 2, 3, 5, 45, 49, 50, 52, 53, 54, 55, 56, 57	10	19
Woodwardia areolata	1, 2, 3, 6, 7, 8, 30, 31, 32, 33, 34, 35, 38	17	33

as well as *Onoclea,* it has been possible to arrive at a percent chemical affinity to the onocleoid complex. These data are given in table 12.

Within the complex itself species of *Matteuccia* have the greatest number of spots, and therefore a higher index of affinity to the complex as a whole. *Onocleopsis* shows the least affinity to the complex with only 37 percent of the total number of compounds; its affinity to *Onoclea* is also very low (28 percent). The relatively high affinity of *Matteuccia struthiopteris* to *Onoclea sensibilis* is significant because this provides supporting evidence of a biochemical nature that these two species are indeed related. If we compare the onocleoid group with all the families investigated, it is apparent that the three genera share more compounds with each other than with any other family.

The chromatographic pattern of *Matteuccia intermedia,* a species of probable hybrid origin between *M. orientalis* and *M. struthiopteris,* shows more similarity to that of *M. orientalis* than of *M. struthiopteris.* Of the twenty-one compounds found in all three species, nine are shared by all. *Matteuccia intermedia* has thirteen spots in common with *M. orientalis* including four not found in *M.*

TABLE 13

Distribution of Compounds in Relation to the Taxa Studied

Taxon	No. compounds restricted to this group	No. compounds in common with onocleoids only	No. compounds in common with onocleoids
Onocleoideae	10	...	...
Aspidiaceae	24	6	17
Blechnaceae	22	2	12
Davalliaceae	13	0	5

struthiopteris (12, 25, 26, 29), and ten in common with *M. struthiopteris* including one spot (1) not found in *M. orientalis.* Only one spot (13) appears in both parent species but not in *M. intermedia.* There are no compounds in the chromatographic pattern of *M. intermedia* that occur in neither parent.

Of the families investigated, there seems to be a greater affinity between the onocleoid complex and Aspidiaceae than to either Blechnaceae or *Davallia.* The onocleoids and Aspidiaceae have six compounds in common which are restricted to these two groups; only two compounds are restricted to the onocleoids and Blechnaceae. Likewise, Aspidiaceae have seventeen compounds in common with the onocleoids whereas Blechnaceae have only twelve. Of all groups investigated, Davalliaceae appear to have the least affinity to the onocleoids.

Among the Aspidiaceous genera investigated, there is a high degree of chemical affinity between *Woodsia* and *Thelypteris* and the onocleoid complex. In fact, the chemical affinity of *Woodsia manchuriensis* and *Thelypteris normalis* to *Onoclea sensibilis* is greater than the affinity of the last species to *Onocleopsis hintonii* or *Matteuccia orientalis.* All but one of the compounds that occur in *Woodsia manchuriensis* occur also in the onocleoid complex.

Genera with very little chemical affinity to the onocleoid complex include *Athyrium* and *Diplazium.*

In summary, few concrete conclusions can be reached concerning the chemical affinities of onocleoid ferns. For the most part, systematic biochemical studies have been restricted to closely related species complexes (Fikenscher and Gibson, 1962; Scora and Wagner, 1964; Smith and Levin, 1963). In many instances comparative chromatographic patterns have yielded profitable indications of affinity between species. At the generic and family level, the diversity in compounds is unknown, and apparently closely related genera may show widely diverse chromatographic patterns. There are indications from my results, however, that (1) onocleoid ferns show greater similarity of compounds among themselves than with other genera, and (2) the collective spot pattern of onocleoid ferns more closely resembles the pattern of genera of Aspidiaceae than that of genera of Blechnaceae.

DISCUSSION AND CONCLUSIONS

Complete analysis of the life-cycle of *Onoclea, Onocleopsis,* and *Matteuccia* has shown that there is a preponderance of similar characteristics in these genera. These include (1) presence of dimorphic fronds, with the fertile frond greatly modified with pinnae tightly inrolled over the sori; (2) graduate sporangia maturation, acropetal in the frond as a whole but basipetal within each sorus; (3) sori arising at the ends of veins, originally distinct but later merging together at maturity; (4) high number of indurated annulus cells in the sporangium; (5) haploid chromosome number of 40 in *Onocleopsis* and *Matteuccia;* (6) green spores; (7) similar perispores; (8) free venation in the fertile frond, remarkably similar in pattern; (9) extrorse membraneous indusium; (10) gametophyte ontogeny; (11) sporophyte development, especially the shape and venation of the earliest fronds; (12) sporophytic trichomes and scales; (13) rhizome anatomy; (14) organization of mechanical tissue in the rhizome; (15) unsclerified and uncutinized sheath around the meristeles in rhizome and stipe; (16) coarse black wiry roots; (17) root anatomy; (18) epidermal cell pattern; (19) structure of palisade and spongy tissues in sterile frond; (20) two-strap vascular bundles in stipe; and (21) chromatographic pattern. Although *Onoclea* and *Matteuccia* differ in such characteristics as chromosome number, venation pattern of the sterile frond, pubescence of the stipe, and prostrate or upright rhizome, these characteristics are found combined in *Onocleopsis.* For example, *Onocleopsis* has the lonchopteroid venation of *Onoclea* as well as the upright rhizome and chromosome number of *Matteuccia.* If it were necessary to choose a prototype from which the two temperate genera evolved, *Onocleopsis* would be the likely choice.

The preponderance of similarities in the three genera, including characteristics that have been considered as important in determining relationships as well as others whose importance is unknown, precludes any possibility that they are unrelated. Although many of their characteristics are found in other genera of Aspidiaceae in no instance do they occur with such high correlation. Therefore, it must be assumed that the onocleoid ferns are closely related to one another and form a separate offshoot of the indusiate ferns.

The twenty-five character states that I have found to differ among the onocleoid, the aspidioid, and the blechnoid ferns are summarized in Table 14. Of these, the onocleoids share twenty-four characters with genera associated by Copeland (1947) with Aspidiaceae, and only four with some species of Blechnaceae. It is readily apparent that: (1) there is an overwhelming morphological similarity between onocleoid ferns and Aspidiaceae; and (2) within the genera of Blechnaceae morphological similarities occur which indicate a high probability of their being related phylogenetically and thereby deserving the familial ranking usually accorded them.

TABLE 14

SUMMARY OF MORPHOLOGICAL FEATURES: COMPARISONS OF FEATURES BETWEEN ONOCLEOIDEAE, ASPIDIACEAE, AND BLECHNACEAE

Onocleoideae	Aspidiaceae	Blechnaceae
1. Spores bilateral	1. Spores bilateral	1. Spores bilateral
2. Perispore present	2. Perispore present	2. Perispore present
3. Spores green	3. Spores non-green	3. Spores non-green
4. Perispore hyaline	4. Perispore hyaline, light brown to variable	4. Perispore reddish-brown or with red-brown base
5. Gametophytic developmental pattern Aspidioid, with no terminal hair	5. Gametophytic developmental pattern Aspidioid, with or without terminal hair	5. Gametophytic developmental pattern Blechnoid, with terminal hair
6. Gametophyte developmental pattern symmetrical	6. Gametophytic developmental pattern symmetrical	6. Gametophytic developmental pattern asymmetrical, angular
7. Gametophytic trichomes unicellular, clavate	7. Gametophytic trichomes unicellular to multicellular	7. Gametophytic trichomes usually bi- to multicellular
8. Basal antheridium cell discoid	8. Basal antheridium cell discoid, rarely broad, distended	8. Basal antheridium cell broad, distended, multicellular; rarely discoid
9. Rhizome dictyostele perforated-pattern	9. Rhizome dictyostele perforated- or Cyatheoid-Stenochlaena-pattern	9. Rhizome dictyostele Cyatheoid- or Stenochlaena-pattern
10. Sclerenchyma: hypodermal ring only	10. Sclerenchyma: hypodermal ring only; also near meristeles	10. Sclerenchyma:rarely hypodermal ring only, also near meristeles and mass in cortex
11. Stipe with two discrete bundles	11. Stipe with two discrete bundles (12 genera), or 3 or more bundles (13 genera)	11. Stipe with two discrete bundles (only *W. areolata*), with 3 or more bundles (all others)
12. Sclerenchymatous bundle sheath absent	12. Bundle sheath absent, partially developed, or completely developed	12. Bundle sheath partially or completely developed; never absent
13. Simple glandular trichomes on fronds	13. Simple glandular trichomes on fronds	13. Simple glandular trichomes on fronds
14. Scales not stiff, erect, tipped with gland or round to elongated cell	14. Scales not stiff, erect, tipped with gland or round to elongate cell, or sometimes stiff, erect with acicular terminal cell	14. Scales stiff, effect, tipped with acicular hair (exc. *W. areolata*)
15. Scale cells broad to elongate, walls thin,	15. Scale cells broad to elongate, walls thin	15. Scale cells narrow, elongate, walls thick,

18. Frond mesophyll cells nondifferentiated; palisade layer nondensely staining or packed	18. Mesophyll cells nondifferentiated; palisade layer nondensely staining	18. Mesophyll cells with densely packed and staining palisade layer, differentiated
19. Fronds dimorphic	19. Fronds mono- to dimorphic	19. Fronds monomorphic except *Blechnum* (*Lomaria*) and *W. areolata*
20. Venation fertile frond free	20. Variable	20. Venation fertile frond usually with costal areole or commissural vein
21. True indusium present	21. True indusium present	21. True indusium absent
22. Indusium membraneous, 1–2 cells thick	22. Indusium membraneous	22. False indusium coarse, 3–5 cells thick
23. True indusium extrorse	23. Indusium introrse to extrorse	23. False indusium introrse
24. Sori terminal on veins	24. Variable	24. Sori above veins
25. Sorus round	25. Sorus round to linear, acrostichoid	25. Sorus linear
26. Sporangia development gradate	26. Sporangia development gradate to mixed	26. Sporangia development gradate
27. Indurated annulus cells 21–45 ($\mathbf{x} = 28$)	27. Indurated annulus cells 10–29 ($\mathbf{x} = 15$)	27. Indurated annulus cells 11–27 ($\mathbf{x} = 18$)
28. Haploid chromosome number 37, 40	28. Haploid chromosome number 27–41	28. Haploid chromosome number 28–37

There are several features which the onocleoids share with one or more species in Blechnaceae. These include the presence of dimorphic fronds, remarkably similar to those found in the *Lomaria* group of *Blechnum;* two vascular bundles in the stipe (only with *Woodwardia areolata*); broad concolorous scales with a round to elongate apical cell (only with *W. areolata*); and reticulate venation (only with *W. areolata*). It is apparent, however, that *Woodwardia areolata* is a highly specialized member of Blechnaceae and many of its characteristics follow a progressive series from more basic elements of Blechnaceae. This is seen in the bi-stranded stipe, which is the termination of a series running from five to three to two strands, the former in species of *Blechnum* and of *Woodwardia* with erect rhizomes. The presence of dimorphic fronds in *W. areolata* is also probably an adaptation to seasonality and does not indicate a relation with the onocleoid ferns. The venation of *W. areolata* is similar to that of *Onoclea* and *Onocleopsis.* Here again, however, one can trace an increasing complexity of venation pattern from mostly free veins in *Blechnum* and some species of *Woodwardia,* to more complex areolation in species of *Woodwardia* but still with free veins runing to the margin of the frond.

The syndrome of characteristics found in Blechnaceae include (1) presence of a false indusium effected by either the inrolled frond margin (as in the *Lomaria* group of *Blechnum*), or by a vestigial structure which probably originated as an inrolled frond margin, later taking on the form of a linear false indusium with a flange forming a broader photosynthetic frond, and finally culminating in a dissected indusium-like structure characteristic of many species of the family today; (2) gametophytic development from a penultimate cell, caused by the appearance of a trichome on the terminal cell on the one-dimensional gametophyte; (3) a reddish-brown or a hyaline perispore with a reddish-brown base; (4) angular gametophytic development, the result of numerous hairs on the margins of the gametophyte and an apparent lateral meristem; (5) bi- or multicellular gametophytic trichomes; (6) a globose to elongate antheridium with a basal cell which is commonly distended or even multicellular; (7) stiff, erect scales on stipe and rhizome, commonly tipped with an acicular hyaline cell; (8) narrow, elongate, thick, dark-walled cells in the scale; (9) sori occurring on top of a commissural vein, which connects two other veins running from the midrib to the margin resulting in a long costal areole in the fertile frond; (10) a densely packed palisade layer in the mesophyll which stains bluish to deep gray, intensely in *Blechnum* and to a lesser extent in *W. areolata;* (11) an introrse false indusium; (12) simultaneous development of sporangia linearly along the vascular commissural vein, rather than circular from the tip of a vein; and (13) a haploid chromosome number which varies from 28 to 35, except for one count of about 37 in one species of *Blechnum.* As most of these characteristics are common to nearly all genera of Copeland's Blechnaceae, including *Blechnum, Salpichlaena, Doodia, Brainea, Sadleria,* and *Woodwardia,* it would be expected that related ferns might also possess most of them. As has been seen, the onocleoids do not.

On the other hand, the syndrome of characters shown by the onocleoids fits well within the diversity found in Aspidiaceae. These include: (1) a true, membraneous, extrorse indusium; (2) a double-stranded vascular system in the stipe; (3) hyaline to light brown perispore without a reddish basal region; (4) symmetrical gametophyte development without a trichome terminating the one-dimensional filament, the gametophyte thus developing symmetrically from an apical meristem; (5) unicellular, clavate trichomes on the gametophyte; (6) discoid basal antheridial cell; (7) typical or perforated dictyostele; (8) sclerenchyma in a hypodermal ring only; (9) no sclerenchymatous bundle sheath in stipe or midrib; (10) scales not stiff, erect, but broader and more lax and pliable (this character varies considerably in the Aspidiaceae, but many genera have scales similar to the onocleoids); (11) mesophyll cells undifferentiated into palisade and spongy layers, and not staining intensely or closely packed; (12) extrorse indusium; (13) sori terminal on veins, arising in a circular pattern rather than linearly; and (14) a haploid chromosome number of 37 or 40. These numbers are rare in Aspidiaceae and 37 has yet to be reported, although *Woodsia* possesses an aneuploid series with numbers of 33, 38, 39, 41; *Dryopteris* has 36 and 41, and *Thelypteris* varies from 27 to 36. It can only be concluded from this study that the onocleoid ferns are related to the great complex of genera placed together in Aspidiaceae.

The affinities of the onocleoids to particular genera of Aspidiaceae are still enigmatic. The lines of relationship within this family are still the object of wide speculation. On the basis of chromosome number *Onoclea* fits nowhere; *Matteuccia* and *Onocleopsis* share the number 40 with *Athyrium, Polystichum, Tectaria,* and *Gymnocarpium. Athyrium* also has a double-stranded vascular structure in the stipe and possesses gametophytic trichomes and other structures similar to the onocleoids. Further elaboration of the relationships of the onocleoids to the Aspidiaceae must await additional studies and more concrete theories on the various phylogenetic lines within that family.

I propose here to combine the onocleoid ferns into a subfamilial group of Aspidiaceae, Onocleoideae, indicating that they are related more closely to one another than to any other group in Aspidiaceae, and that they probably represent a phylogenetic offshoot isolated from any immediate relatives. The evidence for this step is the many characteristics that these three genera share in common. Although many of these are shared individually with other genera of various unrelated affinities within Aspidiaceae, in no instance is there such a high degree of correlation.

TAXONOMY

Family: Aspidiaceae S. F. Gray

Terrestrial or occasionally scandent or epiphytic with a creeping, ascending, or erect rhizome; dictyostelic, paleate; stipe not articulate to the stem; fronds pinnate, simple to decompound, monomorphic or dimorphic; sori dorsal, rarely marginal, circular to elongate to extending indefinitely along the veins or dorsal surface of frond, typically indusiate; spores bilateral, with a conspicuous perispore.

Subfamily: Onocleoideae C. Chr.

Fronds strongly dimorphic, the fertile frond reduced, erect, brown to black when mature; sori enclosed by a laminar structure and indusium; veins free or

anastomosing; sporangia with 17 to 45 indurated annulus cells; spores green.
Genera included: *Matteuccia, Onoclea, Onocleopsis.*

KEY TO GENERA

Venation of sterile frond reticulate:
Rhizome creeping; sterile fronds deltoid-ovate, deciduous, 11–49 cm long..........*Onoclea*
Rhizome erect, trunklike; sterile fronds oblong to elliptic, evergreen, 67–127 cm long....... *Onocleopsis*
Venation of sterile frond free ....*Matteuccia*

Matteuccia Todaro

Matteuccia Todaro, Syn. Pl. Acot. Vasc. Sicilia 30. 1866.
Struthiopteris Willd. Mag. Ges. Nat. Fr. Berlin 3:160. 1809.
Pteretis Raf. Am. Month. Mag. 2:268. 1818.
Pterinodes Siegesb (1736): Kuntze, Rev. Gen. Pl. II:819. 1891.
Pentarhizidium Hayata, Bot. Mag. Tokyo 41:716. 1928.

Terrestrial plants with stout, erect, dictyostelic rhizomes bearing persistent leaf bases. Fronds dimorphic; sterile fronds arising from crown in a tight spiral, 5–30 dm long, 1.6–6 dm broad, deciduous, pinnate with pinnatifid pinnae, the lowermost pinnae diminutive, the venation free; fertile fronds reduced, 28–90 cm long, 2.5–10 cm broad, erect, persistent, pinnate with spreading-ascending, linear to linear-moniliform pinnae, brown to black at maturity. Sori enclosed by the inrolled pinnae. Sporangia with 21–45 indurated annulus cells. Spores bilateral, green, perisporous.

Type species: *Matteuccia struthiopteris* (L.) Todaro (*Osmunda struthiopteris* L.)

KEY TO SPECIES

Pinnae of fertile frond linear-moniliform; sterile frond oblanceolate; pinnae gradually reduced at base; pinnules with 6–11 veins above the midrib, 7–12 below; spores with perispore removed 40–55 microns long ....*M. struthiopteris*
Pinnae of fertile frond linear; sterile frond broadly ovate to linear-elliptic; only the lowermost pairs of pinnae reduced at base; spores with perispore removed 57–95 microns long.
Sterile pinnae 1.7–7.0 mm wide; sterile frond 20–49 cm broad; pinnules with 7–22 veins above the midrib, 7–22 below ....*M. orientalis*
Sterile pinnae 0.8–1.9 mm wide; sterile frond 17–23 cm broad; pinnules with 4–7 veins above the midrib, 4–8 below ....*M. intermedia*

1. *Matteuccia struthiopteris* (L.) Todaro
Syn. Pl. Acot. Vasc. Sicilia 30. 1866.

Osmunda struthiopteris L. Sp. Pl. 2:1066. 1753.
Struthiopteris filicastrum All. Fr. Pedem. 2:283. 1785.
Onoclea struthiopteris (L.) Hoffm. in Hook. and Bak. Syn. Fil. ed. II. 46. 1795.
Onoclea nodulosa Michx. Fl. Bor. Am. ii. 272, 1803.
Struthiopteris germanica Willd. Enum. Pl. Horti. Reg. Bot. Berolin 1071. 1809.
Struthiopteris pensylvanica Willd. Linne Sp. Pl. Ed. Quarto 5:289. 1810.
Onoclea pensylvanica Sm. Rees. Cycl. 23 (3). 1813.
Struthiopteris europaea Hornem. Hort. Reg. Bot. Hafn. 2:945. 1815.
Pterinodes struthiopteris (L.) O. Kuntze, Rev. Gen. Pl. 2. 1891.
Pteretis struthiopteris (L.) Nieuwl. Am. Mid. Nat. 3:197. 1914.
Matteuccia nodulosa (Michx.) Fern. Rhodora 17:164. 1915.
Pteretis pensylvanica (Willd.) Fern. Rhodora 47:123. 1925.
Matteuccia pensylvanica (Willd.) Raymond, Nat. Canad. 77:55–71. 1950.

Description.—Deciduous, medium-sized fern with erect rhizome, with fronds arising from the crown; rhizome 2–9 cm in diameter, 9–20 cm high, heavily covered with persistent frond bases, with black stolons arising from the base and forming new plants; roots black, wiry; fronds dimorphic. Sterile frond: produced March–July, arising in a close spiral from the apex of the rhizome, 50–300 ($\mathbf{x} = 93$) cm in length, 16–60 ($\mathbf{x} = 24$) cm in width, oblong or oblanceolate, pinnate with 30–65 alternate, broadly linear acuminate spreading-ascending pinnae, 9–14 cm in length, 1.2–2.5 cm in width, deeply pinnatifid into oblong blunt segments; pinnae gradually diminutive and clasping toward the base; frond apex pinnatifid, abruptly short-acuminate; venation free from the midrib; petiole grooved, stramineous, dark-brown at the flattened base with thin pale-brown to cinnamon scales, glabrous above. Fertile frond: infrequently produced, July–October, persistent throughout winter, erect, reduced, brownish when mature, narrowly oblanceolate, 30–62 ($\mathbf{x} = 39$) cm in length, 2.5–6 ($\mathbf{x} = 3.2$) cm in width, linear-moniliform with inrolled margins to enclose sori, spreading-ascending; basal pinnae diminutive; stipe grooved; scales as in sterile frond; venation free; sorus with membraneous indusium, sporangia with 21–30 ($\mathbf{x} = 24$) indurated annulus cells, spores 64 per sporangium, bilateral, green, with perispores, 47–73 ($\mathbf{x} = 57$) microns in length with perispore, 40–55 ($\mathbf{x} = 48$) microns in length without perispore; $n = 40$.

Distribution.—Southern China to Japan, Kamchatka, Siberia, west across Russia to Scandinavia and central Europe; North American from Virginia north to Newfoundland, west to British Columbia and Alaska. Isolated populations occur in Florida and the middle western states.

Habitat.—Locally abundant in wet depressions, forming large stands in suitable habitats including flood-plain woods and stream sides; along roadsides in northern portion of its range; frequently found in *Ulmus-Tilia* woodlands in Canada and the northern United States with neutral to slightly calcareous soils, rarely on limestone; near sea level to 4,000 meters elevation.

Specimens examined.—USSR: Siberia, Jenisei, *Lundstrom s.n.* (S); Vicinity of Tobolsk, *Mameev 450* (K), *599* (US). JAPAN: Shinshui Prov. Wada Pass, *Hancock s.n.* (K); Ishikari Prov. *Kudo 4664* (TAI); Honshu Prov. Mt. Fuji, *Hayata s.n.* (TI); Wakayama Prov. Mt. Koya, *Seto 4417* (TNS); Hokkaido Prov. Yunasawa, Hakodate-Shi, *Mutsumoto s.n.* (TI). CHINA: Mongolia centrale, *Licent 3572* (BM); Ku-San, *Hugh s.n.* (BM); Yunnan, *Forrest 14981* (E); Szechuan Prov., between Baor and Tha, *Smith 4878* (S). NORWAY: Hordaland-Fanahd: Totland, *Naustdal s.n.* (BG); SWEDEN: Norrbotten, Tarendo Parish, *Alm s.n.* (UPS); Mt. Snoritjakko, near Lake Tornestrask, *Moldenke & Moldenke 20961.* (NY); Vestrogothia, Paroecia Toarp, *Olson s.n.* (BG); Gastrikland, Gavle, Lovudden, on the bank of river Gaulean, *Nannfeldt 17945* (BG). DENMARK: Killerφd, *Mortensen s.n.* (H); Jylland, Munkebjerg, *Svendsen s.n.* (UPS). GERMANY: Rheinprovinz, *Luersson & Wirtgen 424* (MICH); FRANCE: Les Vosges Mts., *Mougeot s.n.* (NY); CZECHOSLOVAKIA: Kosice, *Suza s.n.* (LE); YUGOSLAVIA: Lower slope of Mt. Triglav, 6 miles from Bled, *Leathes s.n.* (BM). BELGIUM: Flandre orientale, *Hostie s.n.* (BM); IRAN: Ghilan Prov., *Alexeenko s.n.* (MICH). CANADA: Newfoundland, Chimney Cove, *Waghorne, s.n.* (MO); Quebec, St. Anne de la Perado, Portneue Co., *J. & W. Rapp 1105* (MO); Manitoba, Winnipeg, *Denike 1024* (NY); Saskatchewan, 11 miles SW of McKague, *Breitung 478* (US); British Columbia, west of Solitude Mt., on highway from Dawson Creek to Pine Pass, *Calder & Kukkonen 26880* (UPS). UNITED STATES: ALASKA: Talkeetna, *Anderson 7655* (GH); VERMONT: Pittsford, *Slosson s.n.* (NY); CONNECTICUT: Oxford, *Setchell s.n.* (US); NEW YORK: Monroe Co., Henrietta, *Mathews 4390* (UC); NEW JERSEY: Sussex Co., Andover Junction, *Storer 349* (UC); VIRGINIA: Potomac River opposite from Cabin John, Maryland, *Maxon 433* (NY); FLORIDA: Lake Helen, *Noble s.n.* (MO); MICHIGAN: Keweenaw Co., Lake Medora, *Hermann 7948* (MICH); Cheboygen Co., Mud Lake Bog near Douglas Lake, *Ehlers 1620* (UC); INDIANA: La Porte Co., Trail Creek, 2 miles SE of Michigan City, *Tryon 211* (MO); OHIO: Ashtabula Co., Phelps Creek Gorge, *Bartley & Pontius 904* (NY); MISSOURI: Linn Co., Locust Creek Valley, Pershing St. Park, *Steyermark 40475* (GH); Mercer Co., between Moore and Saline, *Palmer and Steyermark 41329* (MO); Livonia, *Bush 7780* (NY); MINNESOTA: Itasca Park, Garrison Point, *Moyle 122* (UC); SOUTH DAKOTA: Rapid City, *Williams & Wilcox s.n.* (NY); Black Hills, Hill City, *Petersen s.n.* (NY); WISCONSIN: Hurley, *Random s.n.* (US); IOWA: Fayette Co., *Fink 258* (US).

Comments.—The North American populations of this species have been recognized as specifically distinct from the European plants. I find no significant differences between plants from these continents and hence maintain them as a single species.

Several forms of the species have been described. These include the following: (1) forma *pubescens* (Terry) Fern., with canescent-tomentose fronds; (2) forma *obtusilobata* (Clute) Fern., with flat, obtusely lobed pinnules; and (3) forma *foliacea* (Farw.) Fern., with the lower half of the fertile frond flat and leaflike, the upper normal. These forms all occur in only a few isolated populations, although they may be frequent among the individuals when found.

2. *Matteuccia orientalis* (Hook.) Trev. Atti 1st. Veneto III. 14:586. 1869.

Struthiopteris orientalis Hook. 2 Cent. Ferns t.4. 1860.
Onoclea orientalis Hook. in Hook. and Bak. Syn. Fil. ed. II, 46. 1862.
Pterinodes orientalis (Hook.) O. Kuntze, Rev. Gen. Pl. 2. 1891.
Struthiopteris cavaleriana Chr. Bull. Ac. Georgr. Bot. Mans. 1904.
Matteuccia cavaleriana (Chr.) C. Chr. Ind. Fil. 420, 1906.
Pentarhizidium japonicum Hayata, Bot. Mag. Tokyo 42:345. 1928.
Pentarhizidium orientalis (Hook.) Hayata. Bot. Mag. Tokyo 42:345. 1928.
Matteuccia japonica (Hay.) C. Chr. Ind. Fil. Suppl. III. 127. 1934.

Description.—Deciduous, coarse, medium-sized with erect, cylindrical rhizome, 15–20 cm long, with fronds arising from the crown; rhizome covered with persistent frond bases and brown lanceolate scales; fronds dimorphic. Sterile fronds produced in spring, arising in a close-spiral, 20–61 cm long, 20–49 cm wide, ovate to ovate-elliptic or broadly lanceolate-deltoid, pinnate with 8–23 pairs of pinnae, opposite below, alternate near the front apex, broadly linear to elliptic-ovate to oblanceolate, spreading-ascending, clasping at base, 5–28 ($\mathbf{x}=14$) cm long, 1.6–7 ($\mathbf{x} = 3.5$) cm wide, deeply pinnatifid into oblong acute lobes; lowermost pair of pinnae diminutive, deflexed; frond apex pinnatifid and abruptly short-acute; venation free from midrib; petiole 14–25 cm long, flattened to grooved above, stramineous, dark-brown and flattened at base, densely covered with linear-lanceolate, brown to cinnamon scales. Fertile fronds produced July–November, persistent throughout winter, reduced, erect, brown-black when mature, oblanceolate to broadly linear, acute to truncate at base, 28–90 cm long, 4.5–7 cm wide, pinnate, with 9–18 pinnae inrolled to enclose sporangia; pinnae 1–10 cm long, 2.5–6 mm wide, linear, ascending; basal pinnae not reduced; stipe terete-grooved, scales as in sterile frond; stramineous to brown when mature; venation free; sorus continuous on both sides of midrib, covered with membraneous indusium; sporangia with 27–45 ($\mathbf{x}=35$) indurated annulus cells; spores 64 per sporangium, bilateral, green, with perispore, 72–110 ($\mathbf{x}=89$) microns long with perispore, 57–95 ($\mathbf{x}=74$) microns long without; $n=40$, 80.

Lectotype.—Hakodadi, Japan, *C. Wright s.n.* (sheet No. H 862/68) (K).

Distribution.—Siberia, Japan, Korea, throughout China to Assam, India and Sikkim.

Habitat.—On wet rocky slopes and by streams in forested areas, 450–4,000 meters elevation.

Specimens examined.—JAPAN: Hokkaido, Yunosawa, Kakodate-Shi, *Matsumoto s.n.* (TI); Tochigi Pres. Nikko Nat'l. Park, Mt. Nyohozan, *Tuyama s.n.* (TI); Osako, Mt. Konyo, *Tagawa 3400,* (TNS); Kyoto, Honshu, interior of Kitashirakawa, *Tagawa 7171* (TNS); Yeso, *de Savatier 1561* (K); Shinshui, Ontake Lake, *Hancock 37* (K); Hakono, *Sasaki s.n.* (TAI). KOREA: Zennan; Mt. Chii, *Bokusawa s.n.* (TNS); CHINA: Kiangsi Prov., Mt. Rosan, *Takahashi 111* (TI); Kweichow, foot of Vanchingshaw, Kiangkow, *Tsiang 7554* (S); Kuling, Kio, *DeVol s.n.* (MICH); Hupeh, Shin Shan Hsien, *Chun & Chien 8289* (UC); Kweichow, near Haikow, Tuyun, *Tsiang 5809* (UC); Yunnan, S. Chungtien, Wu-Tso on Yangtze, *Fen 3297* (GH); Thai-Yong, 60 miles west of port of Swatow, *Dalziel s.n.* (E); Yunnen-sen district, *Cavalerie 1564* (K); INDIA: Assam, Melin, Khasi Hills, *Mann s.n.* (S); Assam, *Simons s.n.* (K); SIKKIM: *Hooker and Thompson s.n.* (K).

Comments.—*Matteuccia orientalis* is variable with regard to size and shape of pinnae. The species consists of two cytological races, a tetraploid found primarily in Japan, and a diploid found on the Asian mainland. The two races appear to differ only in spore size and distribution. The relative uniformity of this species suggests that the tetraploid race may be an autopolyploid.

Japanese pteridologists have suggested that *M. struthiopteris* and *M. orientalis* are phyletically different and have segregated the latter species into a genus of its own, *Pentarhizidium*. This was based on such morphological features as gametophytic structure and development and rhizome anatomy. They have also visualized Japanese material as being different from that of the Asian mainland and have separated it as *P. japonicum* Hayata. The many similarities between *M. struthiopteris* and *M. orientalis* and their apparent successful hybridization to produce *M. intermedia* suggests to me that they are very closely related and should be maintained as a single genus.

The holotype of *Matteuccia cavaleriana* (C. Chr.) C. Chr., based on a single collection by Leveille (E) from the high mountains between Pin-fa and Cou-Yun, China, consists of a fertile frond of *M. orientalis* and a sterile one of *Dryopteris* sp. The young reniform indusia can be seen on the ventral surface of the sterile frond and the vascularization of its stipe consists of 5 to 7 traces, in contrast to the double-stranded condition found in *Matteuccia*. For this reason *M. cavaleriana* is not considered a valid species.

The Wright collection from Japan has been chosen as lectotype because it best displays most readily Hooker's concept of *M. orientalis* and consists of both a sterile and fertile frond.

3. *Matteuccia intermedia* C. Chr.
Filices Purdomianae. Bot. Gaz. 56:337. 1913.

Pentarhizidium intermedium (C. Chr.) Hayata. Bot. Mag. Tokyo 42:346. 1928.
Pteretis intermedia (C. Chr.) Ching, Sunyatsenia 5:244. 1940.

Description.—Deciduous, medium-sized, with erect, cylindrical rhizome, the fronds arising from the crown; fronds dimorphic. Sterile frond: 55–69 cm long, 17–23 cm wide, ovate to broadly lanceolate, pinnate, the pinnae clasping, alternate above and below, opposite centrally, lanceolate to broadly linear, spreading-ascending, 6–17 ($\bar{x} = 10$) cm long, 0.8–1.9 ($\bar{x} = 1.4$) cm wide, pinnatifid one-third to one-half the distance to the midrib into obtuse to dentate lobes; ventral side of pinnae with small, yellow, glandular trichomes on veins; lowermost pinnae diminutive, deflexed; frond apex pinnatifid, the frond base acute-attenuate; venation free from midrib; petiole up to 20 cm long, stramineous, grooved, clothed with scales 2.0–6 mm long, deltoid-attenuate to lanceolate, variable in coloring concolorous tan, concolorous dark-brown, or base or tip dark-brown or tan. Fertile frond: produced April–October, reduced, erect, brown-black when mature, oblanceolate to broadly linear, the base acute to truncate with pinnae somewhat reduced; frond 20–90 ($\bar{x} = 51$) cm long, about 8–10 cm wide, pinnate with pinnules inrolled to enclose sporangia; pinnules 0.5–14 ($\bar{x} = 8$) cm long, 2–6 ($\bar{x} = 3.7$) mm wide, linear, ascending with recurved tips; stipe stramineous to brown with brown base; scales as in sterile frond; venation free, the sorus continuous on both sides of the midrib when mature, indusium wanting; sporangia with 27–34 ($\bar{x} = 31$) indurated annulus cells, spores 64 per sporangium, bilateral, green, with perispore, 60–80 ($\bar{x} = 68$) microns long without perispore.

Lectotype.—*Purdom 8* (S), Shensi, Tai-pei-shan, China.

Distribution.—South-central and southern China to Tibet, Sikkim, and northeastern India.

Habitat.—Mossy banks in deep, shaded forest, 2,000–4,500 meters elevation.

Specimens examined.—CHINA: Yunnan, *Maire 1911* (E); Yunnan, *Delavey 1168* (P);

Montague de Yang, Lankong pres de La-Lie, *Delavey s.n.* (K); S. W. Kansu, Tebbu Country, Wantsang Valley, *Rock 14731* (E, GH, K); Shensi, Tai-pei-shan, *Purdom 1, 8, 9, 12* (G, S, GH); Szechwan, Tienchuan, *Kuan & Wang 2074* (K); Szechwan Prov., Hsu-tsing, *Smith 4804* (S); Tibet Oriental, Prov. de Moupin, *David s.n.* (P). SIKKIM: Between Lachung and Jamukphor, 13000 ft., *Levinge s.n.* (K, E, P); Sikkim, *Bootia 1004* (E). INDIA: Darjeeling, *Cave s.n.* (K, G); Jaunsar District, Lokandi Hill, *Anderson s.n.* (E), *Duthie 19854* (BM, K, S), *Duthie 14465* (OSLO), *Duthie 24717* (K), *Gamble 24898, 14898, 26961* (K).

Comments.—Matteuccia intermedia is somewhat enigmatic because of its apparent rarity in nature, as judged by the low number of collections in herbaria. Many of the characteristics of this species, including length of sterile frond and pinna, form and attitude of basal pinnae, width of fertile pinnae, reduction of basal pinnae of fertile frond, number of indurated annulus cells, and size of spores, are intermediate between those of *M. struthiopteris* and *M. orientalis.* Usually the sterile frond more closely resembles that of *M. struthiopteris* whereas the fertile frond is more similar to that of *M. orientalis.* The intermediate morphological characteristics and the large spores suggest to me that *M. intermedia* is a tetraploid species and probably an allopolyploid derivative of a cross between diploid *M. orientalis* and *M. struthiopteris.*

The specimens of *M. intermedia* that I have examined fall into two morphological categories: (1) those with long, fertile fronds with a rachis 30–66 cm long and pinnae 5–13 cm long and 4–6 mm wide; and (2) those with shorter fertile fronds with a rachis 20–37 cm long and pinnae 3–7 cm long and 2–4 mm wide. Specimens with long fertile fronds have been collected almost entirely in India and Sikkim, but one specimen is known from Yunnan, China. Specimens with short fertile fronds have been collected almost exclusively in China, but one is known from Lokandi Hill, Jaunsar District, India, where both forms occur. The reason for this variation, which is somewhat correlated geographically, is unknown. It may indicate two separate origins of this species by different hybridizations of the same two parental species.

Onoclea L.

Onoclea L. Sp. Pl. 1062. 1753.
Calypterium Bernhardi, Schrad, Jour. 1801:22. 1802.
Riedlea Mirbel. Hist. Nat. des Veg. 5:71. 1803.
Ragiopteris Presl. Tent. Pteridogr. 96. 1836.
Pterinodes O. Kuntze, Rev. Gen. Pl. 2:819. 1891.

Terrestrial plants with cylindrical or somewhat compressed, long-creeping and branching rhizomes, 2–9 mm in diameter, mostly glabrous. Fronds dimorphic; sterile fronds arising from top-half of rhizome, 1.1–4.9 ($\mathbf{x}=2.4$) dm long, 1.3–4.3 ($\mathbf{x}=2.4$) dm broad, deciduous, pinnate with sub-opposite spreading ascending, lanceolate or lanceolate-oblong entire pinnae, the venation reticulate; fertile fronds reduced, erect, persistent, lanceolate-oblong, bipinnate with pinnules rolled into globular divisions, dark brown to black at maturity. Sori enclosed by the inrolled pinnae. Sporangia with 22–40 ($\mathbf{x}=28$) indurated annulus cells. Spores bilateral, green, perisporous.

Type and sole species.—Onoclea sensibilis L.

Onoclea sensibilis L.
Sp. Pl. 1062. 1753.

Onoclea augescens Link. Fil. Sp. 37. 1841.
Calypterium sensibile Bernhardi, Schrad, Jour. 1801. 22. 1802.
Riedlea sensibilis Mirbel. Hist. Nat. des Veg. 5:71. 1803.
Onoclea obtusilobata Schkuhr. Krypto. Gewach. 1:95. 1809.
Ragiopteris obtusilobata Presl. Tent. Pteridogr. 96. 1836.
Ragiopteris onocleoides Presl. Tent. Pteridogr. 96. 1836.
Pterinodes sensibile Kuntze, Rev. Gen. Pl. 2:219. 1891.

Distribution.—Southern Russia, eastern Siberia, southern Kuriles, Mongolia, Manchuria, Japan, Korea, northern China; eastern North America, common from Labrador to Florida and west to Texas; scattered populations west of the Appalachians to Colorado; isolated population in Westmorland, England, probably introduced.

Habitat.—Wet river bottoms and margins of bogs and streams in the southern portion of its range in North America, becoming frequent in wet areas, depressions, and roadsides in northern portion. Sea level to 2,250 meters elevation.

Specimens examined.—USSR: Basin of Klepochnaya River, Littoral Territory, near Voroshilove-Ussuriyskiy, *Boskoba s.n.* (US); District of Vladivostok, west side of Lake Khanka on River Grjaznukha, *Melvil s.n.* (BM); Siberia, *Ehnberg s.n.* (L); Zejskaja Pristan, *Karo 471* (L). KOREA: Diamond Mountain, *L. Gibbs 19* (K); Myongji-San, Kyonggi-Dŏ, *Chung In-cho 4688* (MICH). JAPAN: Hokkaido, Tomakomai Exp. Forest Station, *Y. Kudo 719* (TAI); Honshu, Itabashi, Odawara-Shi, Kanagawa Pres. *H. Ohba s.n.* (TI); Honshu, Dokanyama, Pres. Tokyo, *T. Makino s.n.* (TI); Hokkaido, Shichigorozawa, Kamedagun, Toshima Prov. *Matsumoto s.n.* (TI); Honshu, Tochigi = Senba Kuzuu-mho, Aso-gun, *Tagawa & Iwatsuki 1997* (K); Shinshui, Wada Pass, *Hancock 30* (K); Zenpukuji, Tokyo, Nozu 3, *Lloyd 4323* (UC). CHINA: Manchukuo, Chi-Lin Pr. (Tungpei), *Suzuki-Tokio 9963* (TAI); Mongolia, central, *Licent 3637* (BM); N. Manchuria, *Lude s.n.* (GH). ENGLAND: Westmorland, edge of Lake Elterwater, *Russell H 1318/36* (K); Westmorland, North shore of Rydal Water, Druce's Plant List ed. 2855/1, No. *7208* (K). CANADA: Labrador, island in Hamilton River near Goose Bay, *Gillett & Findley 5689* (US); Newfoundland, Avalon Peninsula, Bay Bulls, *Fernald et al 26154* (US); Newfoundland, Whitbourne, *Robinson & Schrenk 63* (NY); Nova Scotia, near Pictou, *Howe & Lang 508* (NY), Ontario, Paris, collector unknown, 1850–1860 (MICH), Ontario, 4 miles west of Tavistock, Oxford Co. *Soper & Dale 4114* (US). UNITED STATES: MAINE: South Harpswell, *Greeman 3511* (MO); NEW HAMPSHIRE: Hillsboro Co., Temple, *Blake s.n.* (US); NEW YORK: Locust Grove, *Henshaw s.n.* (US); PENNSYLVANIA: Somerset Co., Quemahoning, 1.9 km SSW of Hollsopple, *Shelter s.n.* (NY); VIRGINIA: Giles Co., 1 mile west of Mt. Lake Hotel on road to Cascades, *Wagner 62216* (MICH); NORTH CAROLINA: Swamps near Biltmore, *Biltmore Herbarium No. 683 B* (US); SOUTH CAROLINA: Berkeley Co., Palmer Bridges, 5 miles SE of Honeyhill, *Godfrey & Tryon 133* (US); GEORGIA: Washington Co., Williamson Swamp Creek, 1 mile west of Davisboro, *Tryon & McVaugh 1252* (NY); FLORIDA: Gadsden Co., 1 mile west of Highway 269 on road to Torreya St. Park, *Gillespie 620* (GH); MICHIGAN: Washtenaw Co., Jackson Road, south of Chelsea, *Wagner s.n.* (MICH); INDIANA: Porter Co., Tremont, *Tryon s.n.* (MO); KENTUCKY: Greenup Co., Big Woods, 3 miles from Boyd Co. line, *Smith et al 3548* (US); TENNESSEE: Meriwether Lewis Nat'l. Monument, *King 395* (US); MISSOURI: Ralls County, Oakwood, *Davis s.n.* (MO); ARKANSAS: Hempstead Co., Grassy Lake, *Demaree 46401* (US); MISSISSIPPI: Forrest Co., 8 miles west of New Augusta, *McVaugh 8536* (MICH); LOUISIANA: LaFayette, *DeBaillon s.n.* (NY); Tangipahoa Parish, 1 mile west of Robert, *D. S. & H. B. Correll 9305* (NY); TEXAS: Marshall, *Bush 1007* (MO); Anderson Co., on Neches River at Highway 84, *Whitehouse 1108* (MICH); NORTH DAKOTA: Richland Co., *Stevens 833* (NY); WISCONSIN: St. Croix River Bottom at Osceola, Polk Co., *R. & P. Tryon 4338* (NY); KANSAS: Saline Co., 4 miles east and 4 miles north of Saline, *Humfield 448* (NY); NEBRASKA: Cherry Co., Valentine Lakes Refuge, *Kiener*

23857 (MO); IOWA: Louisa Co., small pond about ½ mile SE of Fredonia, *Davidson 2658* (US); OKLAHOMA: Choctow Co., near Ft. Towson, *Houghton 3957* (MO); COLORADO: Douglas Co., 8 miles SW of Sedalia along highway 67 near Papoose Club, *Weber 4272* (UC).

Fossil record.—Fossils of *Onoclea* have been reported from widely separate areas of the world including Alaska, Canada, United States, central and southern Europe, and Japan. These reports include the following: Bell (1949), *Onoclea hebridica* (Forbes) Bell, from late Cretaceous deposits on Round Island, British Columbia, Canada. Bell also reports Paleocene fossils of this species from western Alberta; Hollick (1906), *Onoclea inquirenda*, specimens remarkably similar in structure to fertile fronds of *O. sensibilis* from Cretaceous deposits in New York; Knowlton (1899), *Onoclea minima* from Laramie Strata of Cretaceous sandstone found in Yellowstone National Park, and *O. neomexicana* from Upper Cretaceous Kirtland Shale, Pina Veta China, San Juan County, New Mexico; Newberry (1898), *Onoclea sensibilis fossilis* from the Tertiary Fort Union group of the Dakatos; Heer (1883), Greenland fossils of *Onoclea;* Hollick (1906), *Onoclea* from the Tertiary floras of Kupreanof Island in southeastern Alaska; Oishi and Huzioka (1941), fossils of *Onoclea* from the Woodwardia Sandstone of Oligocene or upper Eocene in Japan; Forbes (1851), described *Filicites ? hebridica* from Eocene deposits on the Isle of Mull, Scotland; Gardner and Ettingshausen (1882) later placed these in *Onoclea;* and Berry and Hawkins (1935) found *Onoclea* fossils from the Pensauken Formation of New Jersey, a pre-Wisconsin interglacial period in the Pleistocene.

Comments.—Although *O. sensibilis* is more or less constant in morphology throughout its range, various forms have been recognized. These include the following: (1) forma *obtusilobata* (Schkuhr) Gilbert, with sterile fronds bipinnate and intermediate in form between normal sterile fronds and normal fertile fronds, a condition that frequently appears after mowing or grazing although there are reports of its occurrence without previous injury; (2) forma *acutisegmenta* Gruber, with deeply incised or lobed segments; (3) forma *ventroperaferens* Neidorf, with truncate apices and an excurrent rachis in the form of a small straight spine; and (4) forma *hemiphyllodes* (Kiss & Kuemmerle) Weath., with pinnae fertile on one side and sterile on the other.

Onocleopsis F. Ballard

Onocleopis F. Ballard, Am. Fern J. 35:1. 1945.

Terrestrial plants with stout, erect, dictyostelic rhizomes bearing persistent leaf bases. Fronds dimorphic; sterile fronds arising from the crown in a tight spiral, 6.7–12.7 ($\bar{x} = 9.2$) dm long, 2.5–3.2 dm broad, evergreen, pinnate with 17–28 alternate to opposite, lanceolate-oblong, spreading pinnae, truncate or clasping at the base, with serrated dentate, or crenate margins; the lowermost pinnae diminutive, deflexed, the venation reticulate; fertile fronds reduced, 5.6–10.8 ($\bar{x} = 8.2$) dm long, 1.0–1.8 ($\bar{x} = 1.5$) dm broad, erect, not persisting, bipinnate, the pinnules rolled into 3–4 globular divisions, brown at maturity. Sori enclosed by the inrolled pinnae. Sporangia with 17–34 indurated annulus cells. Spores bilateral, green, perisporous.

Type and sole species.—*Onocleopsis hintonii* F. Ballard.

Type collection.—Los Hornos, District of Temascaltepec, Mexico. *Hinton 3297* (K).

Distribution.—Southern Mexico and Northern Guatemala.

Habitat.—In dark, wet, narrow barrancas along streams in muddy soil or on moist clay banks. Pine-Oak Woodland at 2,000–2,800 meters elevation.

Specimens Examined: MEXICO: Mexico, Los Hornos, District of Temescaltepec, *Hinton 3297* (K), *7228* (K, US), *11211* (K, US, UC), *16065* (UC, US), *16314* (UC, US). Morelos, Canyon above Cuernavaca, *Pringle s.n.* (US). Oaxaca, Cerro de San Felipe, 2,100 m, *Conzatti & Gonzalez 480,* (GH); 11.5 miles south of Suchixtepec, Sierra Madre del Sur, *Anderson & Laskowski 4176* (MICH); 7 miles south of Suchixtepec, *Anderson & Laskowski 4166* (MICH); 17.5 miles from Teotitlan del Camino on road to Huautla de Jimenez, *Anderson & Laskowski*

4289. GUATEMALA: Dept. San Marcos, 6 miles south and west of Tajumulco, northwestern slopes of Volcan Tajumulco, *Steyermark 36733* (US).

Comments.—This fern has been collected infrequently and is probably not common anywhere. It should be looked for in dark, narrow canyons between 2,000 and 2,800 meters in the mountainous regions of central and southern Mexico and northern Guatemala. The species is intolerant to frost and cold temperature and its relatively narrow altitudinal distribution may reflect on this factor. Plants occasionally send out short, thick stolons at the base, which will develop into new plantlets. This is similar to the vegetative reproduction in *Matteuccia struthiopteris.* The population north of Teotitlan del Camino consisted mostly of older mature individuals, several feet in height. Populations from the Sierra Madre del Sur were composed of equal numbers of older plants and young sporophytes. The type location at Los Hornos was not found. Hinton reported that much of the original population was wiped out by mining operations and the species may now be extinct at that location.

An additional population of *O. hintonii* was observed by Anderson and Laskowski near the village of Omilteni, about 20 miles west of Chilpancingo, Guerrero, Mexico.

APPENDIX

Voucher Specimens for Morphological Studies

Arachniodes denticulata (Swartz) Ching
COSTA RICA: *Mickel 3221* (NY), stipe

"Aspidium" angulare Kit.
Spores from Rouen, France, botanical garden, no voucher, gametophyte

Asplenium amboinense Willd.
ADMIRALTY ISLANDS: *Grether and Wagner 4139* (UC), chromatography

Athyrium banajaoense Copel.
PHILIPPINES: *Copeland 241* (UC), spores, frond; *Copeland s.n.* (UC), frond, scales

Athyrium crenatum Rupl.
JAPAN: *Nishida s.n.* (UC), spores; *Nishida 60A* (UC), frond, scales

Athyrium nakanoi Makino
TAIWAN: *Chuang and Kao 4182* (UC), chromatography, spores, frond, scales

Athyrium nikkoense Maxon
JAPAN: *Tagawa and Iwatsuki 1953* (UC), chromatography, spores, frond, scales

Athyrium spectabile (Wall) Presl
INDIA: *Hope s.n.* (UC), spores, stipe, frond

Athyrium strigillosum Moore
CHINA: *Tsang 20954* (UC), chromatography, spores, stipe

Athyrium sp.
COSTA RICA: *Mickel 3251* (NY), stipe
AUSTRALIA: *Werner s.n.*, cult. Univ. of California Botanical Garden No. 56.655, gametophytes

Blechnum buchtienii Rosenst.
COSTA RICA: *Mickel 3241* (NY), stipe

Blechnum ensiforme (Liebm.) C. Chr.
MEXICO: *Mexia 9280* (UC), chromatography, spores, scales

Blechnum falciforme (Liebm.) C. Chr.
MEXICO: *Lloyd 4072* (DUKE), roots, frond

Blechnum fraxineum Willd.
MEXICO: *Mexia 9181* (UC), spores, scales
HONDURAS: *Gentle 5071* (UC), chromatography, fronds

Blechnum gibbum (Lab.) Mett.
NEW CALEDONIA: Cult. Univ. of California Botanical Garden No. 62.238-1, sporangia, gametophytes, rhizome, stipe, scales

Blechnum glandulosum Link.
BOLIVIA: Cult. Univ. of California Botanical Garden No. 58.525-S2, gametophyte

Blechnum lehmannii Hier.
COLUMBIA: Cult. Univ. of California Botanical Garden No. 57.206-S1, stipe, frond

Blechnum occidentale L.
MEXICO: *King and Soderstrom 5124* (UC), chromatography, spores, stipe, scales, frond
GUATEMALA: *Weatherwax 183A* (UC), spores, frond
CUBA: *Howard 6146* (UC), frond, scales
JAMAICA: *Crosby, Hespenheide, and Anderson 154* (UC), spores, stipe, frond

Blechnum patersonii (R. Br.) Mett.
Cult. Royal Botanic Garden, Kew, gametophytes

Blechnum polypodioides (Sw.) Trev.
COSTA RICA: *Stork 1534* (UC), chromatography
JAMAICA: *Crosby, Hespenheide, and Anderson 537* (UC), spores, frond, scales

Blechnum procerum Sw.
Cult. Royal Botanic Garden, Kew, gametophytes

Blechnum punctulatum Sw.
Cult. Royal Botanic Garden, Kew, gametophytes

Blechnum sessifolia Broadh.
COSTA RICA: *Rodriquez 110* (UC), chromatography, spores, frond, scales
Blechnum spicant (L.) With.
WASHINGTON: *Rodin 6801* (UC), chromatography, spores, frond
CALIFORNIA: *Rodin 6794* (UC), spores
Blechnum unilaterale Sw.
Cult. Royal Botanic Garden, Kew, gametophytes
Bolbitis donnell-smithii (Christ) Ching
COSTA RICA: *Mickel 2849* (NY), stipe
Bolbitis nicotianifolia (Sw.) Ching
COSTA RICA: *Mickel 2785* (NY), stipe
Brainea insignis Hook.
THAILAND: *Kerr 5925* (UC), chromatography, spores, rhizome, stipe, scales
SUMATRA: *Yates 2652* (UC) spores
Ctenitis acrosora (Hier.) Copel.
COSTA RICA: *Mickel 2975* (NY), stipe
Cyathea sp.
Cult. Univ. of California, Berkeley, gametophytes
Cyclopeltis semicordata (Sw.) J. Sm.
COSTA RICA: No voucher, stipe
Cyclosorus dentata Forsk.
COSTA RICA: *Mickel 2852* (NY), stipe
Cyclosorus patens Sw.
COSTA RICA: *Mickel 3139* (NY), stipe
Cyrtomium falcatum Presl
Cult. Univ. of California Botanical Garden, No. 50.403, gametophytes
Cystopteris fragilis (L.) Bernh.
KASHMIR: *Webster and Nasir 6075* (UC), chromatography, spores
Cystopteris setosa Bedd.
INDIA: *Mann s.n.* (UC), frond
Cystopteris sp.
COSTA RICA: *Mickel 3256* (NY), stipe
Davallia mariesii Moore
JAPAN: *Furuse s.n.* (UC), chromatography
Diacalpe aspidioides Blume
PHILIPPINES: *Quisumbing 2163* (UC), chromatography, spores, scales; *Copeland 161* (UC), spores
INDIA: *Hooker and Thomson s.n.* (UC), spores
Dicksonia squarrosa (Forst.) Sw.
Cult. Univ. of California Botanical Garden, No. 57.1619-1, gametophytes
Didymochlaena truncatula (Sw.) J. Sm.
COSTA RICA: *Mickel 2959* (NY), stipe
Diplazium carnosum Chr.
COSTA RICA: *Stork 437* (UC), spores
Diplazium centripetale (Baker) Maxon
JAMAICA: *Crosby, Hespenheide, and Anderson 547* (UC), chromatography, spores
Diplazium cristatum (Desv.) Alston
COSTA RICA: Cult. Univ. of California Botanical Garden, No. 58.409-3, gametophytes
JAMAICA: *Papenfuss s.n.* (UC), chromatography, spores
Diplazium lonchophyllum Kuntze
Cult. Univ. of California Botanical Garden No. 58.417, frond, sporangia
Diplazium mettenianum (Miq.) C. Chr.
JAPAN: *Togashi s.n.* (UC), chromatography
Diplazium okudaira Makino
JAPAN: *Tagawa and Iwatsuki 3433* (UC), spores

Diplazium polypodioides Bl.
JAVA: Cult. Univ. of California Botanical Garden, No. 58.379-S1, gametophyte
Diplazium unilobum (Poir.) Hier.
JAMAICA: *Crosby, Hespenheide, and Anderson 527* (UC), chromatography, spore
Diplazium werkleanum Christ
HONDURAS: Cult. Univ. of California Botanical Garden No. 57.210-S1, gametophyte
Diplazium wichurae Diels
JAPAN: *Makino s.n.* (UC), chromatography; *Makino 3425* (UC), spores
Diplazium sp.
JAVA: Cult. Univ. of California Botanical Garden, No. 58.361-S1, gametophyte
COSTA RICA: *Mickel 2968, 3010, 3062, 3064* (NY), stipe
Doodia aspera R. Br.
AUSTRALIA: *Tindale s.n.* (UC), spore
Doodia caudata R. Br.
AUSTRALIA: *Clemens s.n.* (UC), chromatography, spore; *Constable s.n.* (UC), spore
Doodia media R. Br.
NEW ZEALAND: *Setchell s.n.* (UC), spore; *Walker 5375* (UC), spore, root, stipe, frond, scale
Dryopteris aemula (Ait.) Kuntze
Cult. Rouen Botanic Garden, France, gametophyte
Dryopteris dentata (Forsk.) C. Chr.
Cult. Univ. of California Botanical Garden No. 58.598-1, stipe, scale
Dryopteris filix-mas (L.) Schott
Cult. Univ. of California Botanical Garden No. 67.174, sporangium, stipe, frond, scale
Dryopteris karwinskyana (Mett.) Kuntze
COSTA RICA: *Mickel 3133* (NY), stipe
Dryopteris kominatoensis Tagawa
JAPAN: *Tagawa 7951* (UC), chromatography, spore
Dryopteris lacera Kuntze
JAPAN: *Furuse s.n.* (UC), chromatography, spore
Dryopteris parallelogramma (Kuntze) Alston
COSTA RICA: *Mickel 3245* (NY), stipe
Dryopteris sabaei C. Chr.
JAPAN: *Suzuki s.n.* (UC), chromatography, spore
Dryopteris setigera (Bl.) Kuntze
JAVA: Cult. Univ. of California Botanical Garden No. 57.771-S1, gametophyte
Dryopteris varia Kuntze
JAPAN: *Furuse s.n.* (UC), spore, scale
Elaphoglossum spp.
COSTA RICA: *Mickel 3019, 3020* (NY), stipe
Faydenia prolifera Hook.
CUBA: *Wright 844* (UC), chromatography, sporangium, spore, frond
Glaphyropteris decussata Presl
COSTA RICA: *Mickel 2973* (NY), stipe
Goniopteris curta Christ
COSTA RICA: *Mickel 2828* (NY), stipe
Goniopteris nicaraguensis Fourn.
COSTA RICA: *Mickel 2764* (NY), stipe
Lastrea hexagonoptera (Michx.) C. Chr.
PENNSYLVANIA: *L.S. s.n.* (UC), spore
Lastrea mercurii Braun
COSTA RICA: *Mickel 3138c* (NY), stipe
Lastrea rudis Kuntze
COSTA RICA: *Mickel 3013* (NY), stipe

Lastreopsis exculta (Mett.) Tindale ssp. *Guatemalensis* (Baker) Tindale
COSTA RICA: *Mickel 3057* (NY), stipe
Lomariopsis sp.
COSTA RICA: *Mickel 2792* (NY), stipe
Marginariopsis wiesbaurii (Sod.) C. Chr.
COSTA RICA: *Lloyd 4254* (UC), spore, germination
Matteuccia intermedia C. Chr.
CHINA: *Delavey 1168* (P), frond, scale; *Kuan and Wang 2074,* (K), frond, scale; *Purdom 1* (GH), chromatography, frond, stipe, scale; *Purdom 12* (GH), frond; *Rock 14731* (GH), chromatography, spore, stipe, frond, scale; *Smith 4804* (UPS), spore, sporangium, stipe, frond.
SIKKIM: *Levinge s.n.* (K), sporangium, spore, stipe, frond
TIBET: *David s.n.* (P), frond, scale
INDIA: *Cave s.n.* (K), sporangium, frond, scale; *Gamble 15812* (K), frond
Matteuccia orientalis (Hook.) Trev.
JAPAN: *Albrecht s.n.* (LE), frond; *Brooks 619* (UC), spore; *Copeland 2807* (MICH) spore; *Faurie 5615* (MICH), spore; *Holbrook s.n.* (NY), spore; *Lloyd 4317* (UC) from Nozu A, cult. Univ. of California Botanical Garden, spore, rhizome, root, stipe, frond, scale, gametophyte, cytology, chromatography; *Lloyd 4337* (UC) from Nozu A, gametophytes; *Shiota 1511* (GH), spore; *Stanford s.n.* (NY), spore; *Tagawa 7171* (GH), spore; *Wright s.n.* (K), spore, sporangium, frond, stipe, scale
KOREA: *Taquet 47* (MICH) spore
CHINA: *Faber s.n.* (NY), spore; *Henry 5113* (US), spore; *Maire s.n.* (GH), spore; *Rock 7569* (US, NY), chromatography, spore; *Steward, Chiao, and Cheo 448* (GH), chromatography, spore; *Takahashi 111* (TI), frond; *Tsiang 5809* (NY), spore
INDIA: *Clarke 42623* (NY), spore; *Mann s.n.* (US), chromatography, spore; *Taquet 3643* (UC), spore
Matteuccia struthiopteris (L.) Tod.
ALASKA: *Calder 6542* (UC), chromatography
BRITISH COLUMBIA: *Calder and Kukkonen 26880* (UPS), frond
CANADA: *Cody and Spicer 11821* (UC), chromatography
CONNECTICUT: *Lloyd 4307* (UC), spore, gametophyte
MICHIGAN: *Lloyd 4309* (UC), chromatography, root, rhizome, stipe, frond, gametophyte; *Lloyd 4311* (UC), chromatography, spore, root, rhizome, stipe, frond, scale, gametophyte
FINLAND: *Alava 2599* (UC), chromatography
NORWAY: *Kelano s.n.* (UC), chromatography
SWEDEN: *Bengtsson s.n.* (S), frond; *Enander s.n.* (S), frond, scale; *Lagerkranz s.n.* (S), frond; *Moldenke and Moldenke 20961* (NY), chromatography
SCANDINAVIA: *Heikkinen s.n.* (UC), chromatography; *Thedenius s.n.* (NY), chromatography
JAPAN: *Brooks 616* (UC), chromatography; *Faurie 5616* (MICH) chromatography, spore; *Lloyd 4327* (UC) from Nozu C, cult. Univ. of California Botanical Garden, spore, root, rhizome, stipe, frond, scale, gametophyte; *Lloyd 4341* (UC) from Nozu A, spore, root, rhizome, stipe, frond, scale, gametophyte; *Tagawa 6321* (US), spore
RUSSIA: *Karelin and Kiriloff 1097* (NY), spore; *Mameev 599* (US), chromatography
Meniscium serratum Cav.
COSTA RICA: *Mickel 2769* (NY), stipe
Nephrolepis biserrata (Sw.) Sch.
SUMATRA: *Toroes 179* (UC), chromatography
Nephrolepis cordifolia Presl
JAPAN: *Togasi 1179* (UC), stipe, scale
Nephrolepis occidentalis Kuntze
COSTA RICA: No voucher, stipe
Oleandra sp.
COSTA RICA: *Mickel 3124* (NY), stipe

Onoclea sensibilis L.

Connecticut: *Lloyd 4303* (UC), spore; *Lloyd 4304* (UC), spore
Florida: *Lloyd 4297* (UC), chromatography, spore
Georgia: *Lloyd 4298* (UC), chromatography, spore
Maryland: *Lloyd 4302* (UC), chromatography, spore
Michigan: *Lloyd 4312* (UC), chromatography, spore, root, rhizome, stipe, frond, scale, gametophyte; *Lloyd 4426* (UC), from Knobloch A, cult. Univ. of California Botanical Garden, spore, root, rhizome, stipe, frond, scale, cytology; *Wagner s.n.* (MICH), frond, scale.
New Hampshire: *Lloyd 4306* (UC), chromatography, spore; *Lloyd 4305* (UC), spore
New York: *Rodin 6657,* cult. Univ. of California Botanical Garden, spore, gametophyte
North Carolina: *Lloyd 4301* (UC), chromatography, spore
Ohio: *Lloyd 4308* (UC), chromatography, spore
South Carolina: *Lloyd 4300* (UC), chromatography
Texas: *Lloyd 4294* (UC), chromatography, spore, root, rhizome, stipe, frond, scale
JAPAN: *Lloyd 4323* (UC), from Nozu B, cult. Univ. of California Botanical Garden, chromatography, frond, scale, gametophyte; *Lloyd 4425* (UC), from Nozu A, cult. Univ. of California Botanical Garden, root, rhizome, stipe, frond, gametophye; *Makino 34278* (UC), spore; *Masamune s.n.* (NY), chromatography; *Tagawa and Iwatsuki 1997* (US), chromatography; *Watanabe s.n.* (GH), spore
KOREA: *Smith s.n.* (US), chromatography
MANCHURIA: *Lude s.n.* (GH), spore
RUSSIA: *Karo 471* (NY), spore

Onocleopsis hintonii F. Ballard

MEXICO: *Anderson and Laskowski 4176* (MICH), spore, gametophyte; *Anderson and Laskowski 4289* (MICH), spore, root, rhizome, stipe, frond, gametophyte; *Lloyd 4082* (UC), spore, root, rhizome, stipe, frond, scale, gametophyte; *Lloyd 4084A* (UC), spore, root, rhizome, stipe, frond, scale, gametophye; *Lloyd 4313* (UC), from Anderson and Laskowski B, cult. Univ. of California Botanical Garden, chromatography, spore, root, rhizome, stipe, frond, scale, gametophyte; *Lloyd 4314* (UC), from Anderson and Laskowski plant G, chromatography, root, rhizome, stipe, frond, scale, spore, gametophyte; *Lloyd 4315* (UC) from Anderson and Laskowski plant A, chromatography, spore, root, rhizome, stipe, frond, scale, gametophyte, cytology; *Lloyd 4331* (UC) from Anderson and Laskowski plant D, chromatography, spore, gametophyte.

Peranema cyatheoides Don

INDIA: *Mann s.n.* (UC), chromatography, spore; *Wallich 183* (UC), spore, stipe, frond, scale

Peranema formosana Hayata

TAIWAN: *Kanehira and Sasaki 21671* (UC), chromatography; *Kanehira and Sasaki s.n.* (UC), spore; *Fruzie s.n.* (UC), spore

Plecosorus speciosissimus (A. Br.) Moore

COSTA RICA: *Mickel 3247* (NY), stipe

Polybotrya cervina (L.) Kaulf.

COSTA RICA: *Mickel 3043* (NY), stipe

Polybotrya osmundacea HBW

COSTA RICA: No voucher

Polystichum munitum Presl

California: *Hooker 9050* (UC), spore
Oregon: *Rodin 6933* (UC), chromatography, spore

Polystichum retroso-paleaceum Tag.

JAPAN: *Togasi s.n.* (UC), chromatography, spore, stipe, frond, scale

Polystichum platyphyllum (Willd.) Presl

COSTA RICA: *Mickel 3230* (NY), stipe

Pteris quadriaurita Retz.

Cult. Univ. of California Botanical Garden No. 60.1162-S1, spore germination

Sadleria cyatheoides Kaulf.
HAWAII: *Fosberg 9363* (UC), chromatography, spore; *Grether 4281* (UC), spore
Sadleria polystichoides (Brack.) Heller
HAWAII: *Grether and Wagner 3924* (UC), spore, frond, stipe, scale
Sadleria sp. (= *S. intermedia*)
HAWAII: *Faurie 39* (UC), chromatography, spore
Steiropteris brachyodus Kuntze
COSTA RICA: *Mickel 2847* (NY), stipe
Tectaria cicutaria (L.) Copel.
Cult. Botanischer Garten, Munchen-Nymphenburg, gametophyte
Tectaria euryloba (Christ) Maxon
COSTA RICA: *Mickel 2788* (NY), stipe
Tectaria melanocaulis (Bl.) Copel.
JAVA: Cult. Univ. of California Botanical Garden No. 55.760, stipe, frond, scale
Tectaria plantaginea (Jacq.) Maxon
COSTA RICA: *Mickel 2787* (NY), stipe
Thelypteris normalis Moxley
JAMAICA: *Orcutt 5998* (UC), chromatography, spore
HAITI: *Leonard and Leonard 13630* (UC), spore
Thelypteris torresiana (Guad.) Alston
COSTA RICA: *Mickel 3059* (NY), stipe
Woodsia ilvensis (L.) R. Br.
CANADA: *Baxter s.n.* (UC), scale
PENNSYLVANIA: *Heller s.n.* (UC), chromatography, spore
Woodsia manchuriensis Hook.
JAPAN: *Furuse s.n.* (UC), chromatography, stipe, frond, root; *Tagawa and Iwatsuki 1951* (UC), stipe, frond, scale, spore
Woodsia mexicana Fee
MEXICO: *Correll and Gentry 2286* (UC), spore
Woodsia polystichoides Eaton
JAPAN: *Furuse s.n.* (UC), spore, stipe, frond, scale
Woodwardia areolata (L.) Moore
FLORIDA: *Clewell 2390*, cult. Univ. of California Botanical Garden, chromatography, spore, root, rhizome, stipe, frond, scale; *Lloyd 4296* (UC), chromatography, scale, spore
MISSISSIPPI: *Lloyd 4395* (UC), spore
NEW JERSEY: *Storer 371* (UC), chromatography, spore, scale, stipe, frond
SOUTH CAROLINA: *Clausen and Trapido 3667* (UC), spore, stipe, frond, scale

LITERATURE CITED

ALSTON, A. H. G.

1956. The subdivision of the Polypodiaceae. Taxon 5:23–25.

ATKINSON, L. R., and A. G. STOKEY

1964. Comparative morphology of the gametophyte of homosporous ferns. Phytomorphology 14:51–70.

BEEKS, R. M.

1955. Improvements in the squash technique for plant chromosomes. Aliso 3:131–133.

BELL, W. A.

1949. Uppermost Cretaceous and Paleocene floras of western Alberta. Can. Dept. Mines and Tech. Surv. Geol. Surv. Bull. 13:1–231.

BERRY, E. W., and A. C. HAWKINS

1935. Flora of the Pensauken formation in New Jersey. Bull. Geol. Soc. Am. 46:245–251.

BOWER, F. O.

1908. The origin of a land flora. MacMillan, London.

1914. Studies in the phylogeny of the Filicales, IV. *Blechnum* and allied genera. Ann. Bot. 28:363–431.

1923. The Ferns (Filicales) treated comparatively with a view to their natural classification. Vol. 1. Analytical examination of the criteria of comparison. Cambridge Univ. Press, England.

1928. *Ibid,* vol. 111. The leptosporangiate ferns. Cambridge Univ. Press, England.

CAMPBELL, D. H.

1905. The structure and development of mosses and ferns. Macmillan, New York.

CHIARUGI, A.

1960. Tavole cromosomiche delle Pteridophyta. Caryologia 13:27–150.

CHING, R. C.

1940. On natural classification of the family "Polypodiaceae." Sunyatsenia 5:201–268.

CHRISTENSEN, C.

1906. Index Filicum. Copenhagen.

1938. Filicinae. *In* Manual of Pteridology (Fr. Verdoorn, ed.), The Hague, Netherlands.

COPELAND, E. B.

1947. Genera Filicum. Ronald Press, New York.

DIELS, L.

1902. Polypodiaceae. *In* A. Engler and K. Prantl. Die naturlichen Pflanzenfamilien, Bd. 1 (Abt.) 4:139–339. Leipzig.

FABBRI, F.

1963. Primo supplemento alle Tavole cromosomiche delle Pteridophyta di Alberto Chiarugi. Caryologia 16:237–335.

1965. Secondo supplemento alle Tavole cromosomiche delle Pteridophyta di Alberto Chiarugi. Caryologia 18:675–731.

FIKENSCHER, L. H., and M. R. GIBSON

1962. Investigations of *Dryopteris* species. 111. The occurrence of phloroglucinol derivatives in various American species of *Dryopteris.* Lloydia 25:196–200.

1851. Note on the vegetative remains from Ardtun Head. Quart. J. Geol. Soc. Lond. VII.

FORBES, E.

FOSTER, A. S.

1966. Morphology of anastomoses in the dichotomous venation of *Circaeaster*. Am. J. Bot. 53:588–599.

1968. Further morphological studies on anastomoses in the dichotomous venation of *Circaeaster.* J. Arnold Arb. 49:52–67.

GARDNER, J. S., and C. B. ETTINGSHAUSEN

1882. British Eocene flora, vol. 1. Pal. Soc. Lond.

GOEBEL, K.

1898. Organographie der Pflanzen insbesondere der Archegoniaten und Samenpflanzen. Jena.

GREGORY, B. A.
1904. The reduction division in ferns. Proc. Roy. Soc. Lond. B. 73:86.

HANNIG, E.
1911. Uber das Vorkommen von Perisporien bei den Filicinen nebst Bemerkungen uber die systematische Bedeutung derselben. Flora 103:321–346.

HARA, N.
1964. Ontogeny of the reticulate venation in the pinna of *Onoclea sensibilis*. Bot. Mag. Tokyo 77:381–387.

HAYATA, B.
1928. On the systematic importance of the stelar system in the Filicales, III. Bot. Mag. Tokyo 42:334–348.

HEER, O.
1883. Die fossile Flora Groenlands, II. Die Tertiare flora von Groenland. Flora Fossilis Artica, ed. VII.

HOLLICK, A.
1906. The Cretaceous flora of southern New York and New England. U. S. Geol. Surv. Mon. 50.

HOLTTUM, R.
1947. A revised classification of the leptosporangiate ferns. J. Linn. Soc. (Bot.) 53:123–158.
1949. The classification of ferns. Biol. Rev. 24:267–296.
1965. The British ferns in a world perspective. Brit. Fern Gaz. 9:205–212.

HOOKER, W. J.
1874. Synopsis Filicum. Hardwicke, London.

KAWASAKI, T.
1956. Studies on the sexual generation of ferns (9). On the prothallia of some species of *Rumohra* and *Cyrtomium*. J. Jap. Bot. 31:333–339.

KEATING, R. C.
1968. Trends of specialization in the stipe anatomy of *Dennstaedtia* and related genera. Am. Fern J. 58:126–150.

KLEKOWSKI, E. J., JR.
1968. Reproductive biology and evolution in the Pteridophyta. Ph.D. thesis. Univ. of California, Berkeley.
1969. Reproductive biology of the Pteridophyta, III. The Blechnaceae. J. Linn. Soc. (Bot.) 62:360–377.

KLEKOWSKI, E. J., JR., and R. M. LLOYD
1968. Reproductive biology of the Pteridophyta, 1. General considerations and a study of *Onoclea sensibilis*. L. J. Linn. Soc. (Bot.) 60:315–324.

KNOWLTON, F. H.
1899. Fossil floras of the Yellowstone National Park. U. S. Geol. Surv. Mon. 32:651–882.

LANJOUW, J., and F. A. STAFLEU
1964. Index Herbariorum. part 1. The herbaria of the world. 5th ed. Regnum Veg. 31, Utrecht.

LITARDIERE, R. DE
1921. Recherches sur l'element chromosomique dans la caryocinese somatique des Filicinees. La Cellule 31:255–473.

LLOYD, R. M.
1965. Clinal patterns in frond anatomy of *Polypodium*. Madroño 18:65–74.
1969. Occurrence of a winged perispore in a New World species of *Polypodium*. Brittonia 21:80–82.
1970. A survey of some morphological features of the genus *Elaphoglossum* in Costa Rica. Am. Fern J. 60:73–83.

LLOYD, R. M., and E. J. KLEKOWSKI, JR.
1970. Spore germination and viability in Pteridophyta: Evolutionary significance of chlorophyllous spores. Biotropica 2:129–137.

MARENGO, N. P.
1949. A study of cytoplasmic inclusions during sporogenesis in *Onoclea sensibilis.* Am. J. Bot. 36:603–613.
1956. The microscopic structure of the mature spores of the sensitive fern, the ostrich fern, and the royal fern. Am. Fern J. 46:97–104.

MEHRA, P. N.
1949. Some observations on the prothallus of *Diplazium esculentum* (Retz) Sw. J. Indian Bot. Soc. 23:1–8.

MOMOSE, S.
1937*a*. Studies on the gametophytes of ferns (I). On the prothallial development and prothallium of *Dryopteris varia* O. Kuntze. J. Jap. Bot. 13:13–120.
1937*b*. *Ibid* (II). On the prothallium of *Cyrtomium falcatum* (L. fil) Presl and *Rumohra standishii* (Moore) Nakai. J. Jap. Bot. 13:414–424.
1938*a*. *Ibid.* (III). On the prothallium of *Polystichum aculeatum* Roth. var *taiwanense* Nakai & Momose and *Leptogramma totta* J. Smith. J. Jap. Bot. 14:56–62.
1938*b*. *Ibid.* (IV). On the prothallium of *Diplazium thunbergii* Nakai and *Diplazium latifolium* Moore, J. Jap. Bot. 13:265–273.
1938*c*. *Ibid.* (V). On the prothallium of Eudryopteroid ferns. J. Jap. Bot. 14:407–420.
1938*d*. *Ibid.* (VII). On the prothallium of some athyrioid ferns. J. Jap. Bot. 14:532–540.
1939. *Ibid.* (XII). On the prothallium of athyrioid ferns. J. Jap. Bot. 15:712–726.
1940. *Ibid* (XIII). J. Jap. Bot. 16:206–215.
1958. Prothallia of onocleoid ferns. J. Jap. Bot. 33:289–296.

MOORE, T.
1857. Index Filicum. Williams and Norgate, London.

NAYAR, B. K.
1960. The gametophyte and young sporophyte of *Athyrium esculentum.* Am. Fern J. 50: 194–203.
1961. Morphology of the gametophyte and young sporophyte of *Matteuccia orientalis* (Hk.) Trev. J. Indian Bot. Soc. 40:502–510.
1962. Gametophytes of some species of *Blechnum.* J. Indian Bot. Soc. 41:33–44.
1967. Morphology of the spores and prothallus of *Christiopteris tricuspis.* Am. Fern J. 57:15–27.
1968. The prothallus of *Matteuccia pensylvanica.* Brit. Fern Gaz. 10:26–29.

NAYAR, B. K., N. BAJPAI, and F. RAZA
1966. Morphological studies on some species of *Blechnum, Doodia, Woodwardia,* and *Stenochlaena:* 1. The gametophytes and the juvenile sporophytes. J. Linn. Soc. (Bot.) 59:405–423.

NAYAR, B. K., and P. CHANDRA
1965. The gametophyte of some species of *Lastrea* Bory. J. Indian Bot. Soc. 44:84–94.

NAYAR, B. K., and S. DEVI
1964. Spore morphology of Indian ferns. 1. Aspidiaceae. Grana Palynol. 5:80–120.

NAYAR, B. K., and S. KAUR
1964. Contributions to the morphology of *Tectaria:* the spore, prothalli, and juvenile sporophytes. Bull. Torrey Bot. Club. 91:95–105.
1966. On the fern genera *Acrophorus, Diacalpe, Lithostegia* and *Peranema.* Bot. Notiser 119:1–23.

NAYAR, B. K., and F. KAZMI
1963. Ferns of India. X. *Matteuccia.* Bull. Natl. Bot. Gard. 82:1–16.

NEWBERRY, J. S.
1898. The later extinct floras of North America. U. S. Geol. Surv. Mon. 35.

OGURA, Y.
1938. Anatomie der Vegetationsorgane der Pteridophyten *in* Handbuch der Pflanzenanatomie (Linsbauer, ed.) Abt. 2, 7 (Teil 2): i–viii, 1–476. Berlin.

OISHI, S., and Z. HUZIOKA

1941. Studies in the Cenozoic plants of Hokkaido and Karahuto. 1. Ferns from the Woodwardia Sandstone of Hokkaido. J. Fac. Sci. Hokkaido Imp. Univ. Ser. 4, Geol. and Min. 6:177–192.

ORTH, R.

1938. Zur Morphologie der Primarblatter einheimischer Farne. Flora 133 (N.F. 33):1–55.

PICHI-SERMOLLI, R. E. G.

1958. The higher taxa of the Pteridophyta and their classification *in* Systematics of Today (O. Hedberg, ed.). Uppsala.

RAO, A. R., and P. KHARE

1964. Contributions to our knowledge of the sporophyte of *Tectaria amplifolia* (V.A.V.R.) Christensen. Proc. Indian Acad. Sci. 59:329–339.

SCHOUTE, J. C.

1938. Anatomy. *In* Manual of Pteridology (Fr. Verdoorn, ed.), Nijhoff, The Hague, Netherlands.

SCORA, R. W., and W. H. WAGNER, JR.

1964. A preliminary chromatographic study of eastern American *Dryopteris*. Am. Fern J. 54:105–113.

SLOSSON, M.

1906. How ferns grow. Holt, New York.

SMITH, D. M., and D. A. LEVIN

1963. A chromatographic study of reticulate evolution in the Appalachian *Asplenium* complex. Am. J. Bot. 50:952–958.

STOKEY, A. G.

1950. The contribution of the gametophyte to classification of the homosporous ferns. Phytomorphology 1:1–20.

STOKEY, A. G., and L. R. ATKINSON

1952*a*. The gametophyte of *Stenochlaena palustris* (Burm.) Bedd. Phytomorphology 2:1–9.

1952*b*. The gametophyte of *Blechnum spicant* (L.) Wither. and *B. buchtienii* Rosenst. Phytomorphology 2:9–15.

1957. The gametophyte of some American species of *Elaphoglossum* and *Rhipidopteris*. Phytomorphology 7:275–292.

STONE, I. G.

1961. The gametophytes of the Victorian Blechnaceae. 1. *Blechnum nudum* (LaBill.) Luerss. Aust. J. Bot. 9:20–36.

TSCHUDY, R. H., and B. D. TSCHUDY

1965. Modern fern spores of Rancho Grande, Venezuela. Acta Bot. Venezuelica 1:9–71.

WAGNER, W. H., JR.

1952. The fern genus *Diellia*. Univ. of California Pub. Bot. 26:1–212.

1954. Reticulate evolution in the Appalachian Aspleniums. Evolution 8:103–118.

1969. The construction of a classification. *In* Systematic Biology, Natl. Acad. Sci., Washington, D.C.

PLATES

Onoclea sensibilis, habitat. *a.* Plants in fully exposed habitat in wet marshy area ca. 22 miles west of Allendale, South Carolina, on US 301, in Georgia. September 1967. *b.* Plants beneath bridge at same location. Fronds ca. 1 meter tall.

a

b

Matteuccia struthiopteris. a. Habitat, in low depressions under *Acer rubrum,* Matthaei Botanical Garden, University of Michigan, September 1967. *b.* Plant cultivated at University of California Botanical Garden.

Onocleopsis hintonii. *a*. In dense shade in barranca, 17.5 miles east of Teotitlan del Camino on road to Huautla de Jimenez, Oaxaca, Mexico. February 1967. *b*. Plant cultivated at University of California Botanical Garden.

Type specimen of *Onocleopsis hintonii*. *a*. Sterile frond. *b*. Fertile frond.

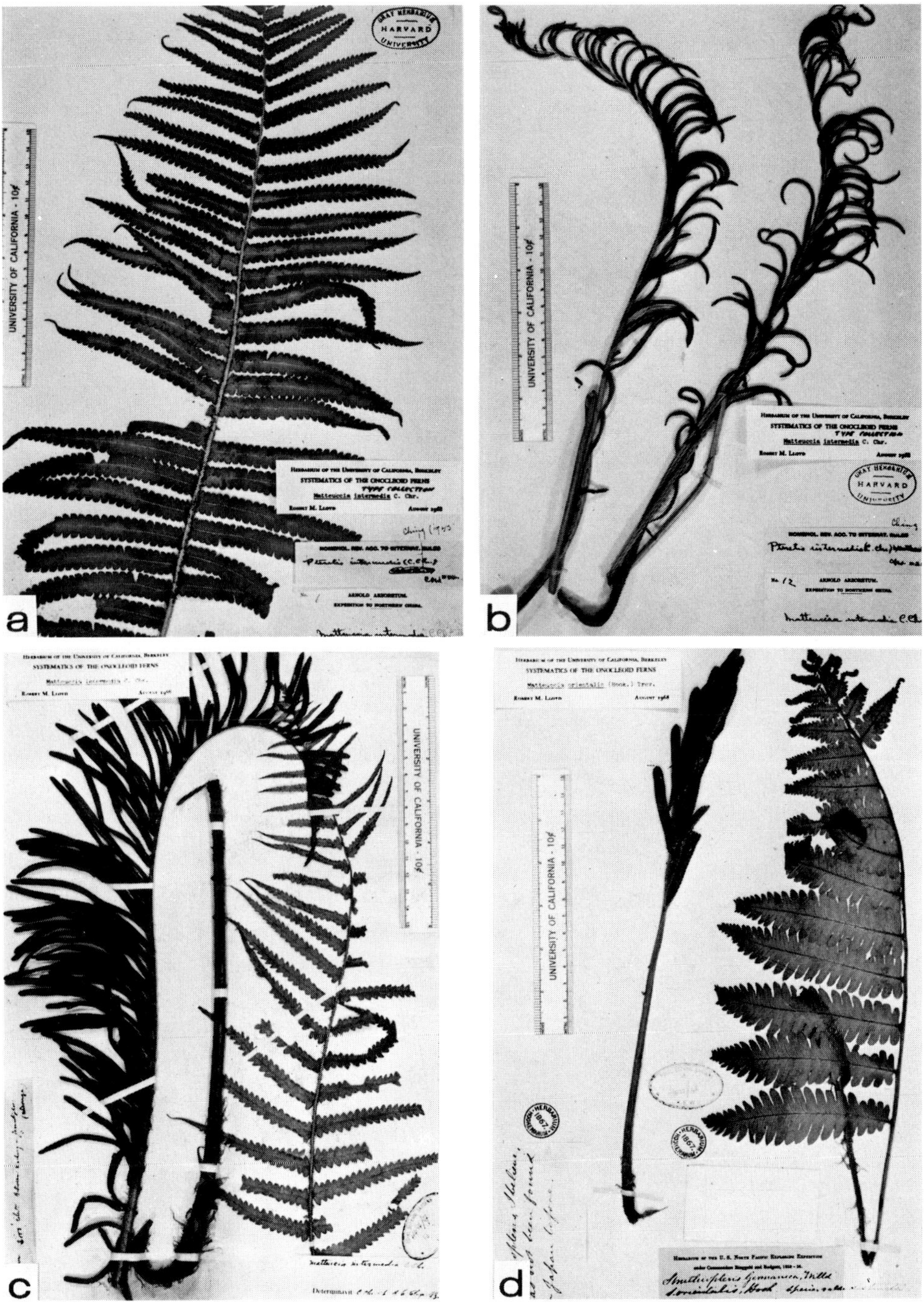

Herbarium specimens of *Matteuccia*. *a*. Isotype of *M. intermedia,* sterile frond, collected at Shensi, Tai-pei-shan, China, *Purdom 1* (GH). *b*. Isotype of *M. intermedia,* fertile frond. *c*. *M. intermedia,* typical plant with large fertile frond and pinnae, from Sikkim. *d*. Lectotype of *M. orientalis* from Hakodadi, Japan, *C. Wright s.n.* (K).